# The Path to Everywhere and Nowhere

## *The Trouble with a Unification*

***First Edition***

The Path to Everywhere and Nowhere: The Trouble with a Unification.



Copyright Office Registration Number: TXu-1-981-382

ISBN: 978-1-7368598-5-8

Other books by The Heart of Man can be found at:

The Heart of Man Publishing - Your Source for Anonymous Publishing (www.roadtoatoe.net):

**The Resurrection of Man**: The Story of Adam and Eve Told Mathematically.

**Logic's Dilemma**: What is Truth?

# Table of Contents

# Message to the Reader

In this book, we analyze, from a self-taught perspective, the question, "Will there be a fundamental theory of the laws that govern the world we live in?" Shrouded in extreme abstraction embodied in self-described gobbledygook, accessible only to a trained practitioner, modern physics can seem like a capricious subject. Nevertheless, with perseverance, at least a rudimentary understanding of modern physics is available to the uninitiated spectator. This book is a testament to what can be learned from the outside looking in. The hope is that others will appreciate the simplicity of the ideas contained within modern physics once all the jargon is clarified.

The first Chapter examines where we are at the moment and reveals what modern physics tells us about our world. Why do so many physicists believe that a comprehensive theory is possible? And why is it so hard to uncover such a theory? Next, in Chapter 2, we examine the struggle physicists have endured attempting to uncover a comprehensive theory. The answer lies in the sequence of discoveries that drove physics to where it is today. It is well known that the two theories that dominate physics, the general theory of relativity and the quantum theory, are incompatible. The search for a comprehensive theory has revolved around finding an approach that unifies the two theories in a manner that will overcome their apparent inherent incompatibilities. Unfortunately, the most promising approach, string theory, has so far failed to live up to its potential. Moreover, sophisticated collider experiments have failed to uncover evidence of the hypothesized structures, such as supersymmetry, that would indicate which direction the search for a comprehensive theory should turn.

In Chapter 3, we examine what our current theories tell us about our world. They tell us much but also reveal gaps in our understanding that require solutions. Various approaches attempt to go beyond the revelations in the standard particle physics model. Those approaches will be examined in detail. But finding a successful approach has been challenging, primarily because of the conceptual issues that have prevented uncovering a reasonable unified solution. Such issues are discussed in detail in Chapter 4. Finally, searching for a comprehensive theory has driven physicists into regions beyond the traditional boundaries of physics and into the world of speculative philosophy. In this environment, not all physicists feel secure. Yet, searching for a unified theory raises age-old philosophical questions that the rationalist and empiricist philosophers of past generations struggled to answer. Do we have better answers today? A discussion of those questions appears in Chapter 5.

The standard model of particle physics attempts to describe all fundamental forces and the relationships between elementary particles using a single theoretical framework. In physics, forces are described by fields that mediate interactions between separate objects. The fields are conceived as existing at all points of space and time, in essence, of infinite duration. Quantizing the fields turns the infinite spread of the fields into a smaller number of discrete entities (particles). James Clerk Maxwell formulated the first field theory of electromagnetism, where a force field surrounds an electrified particle. Then Albert Einstein developed general relativity, a field theory of gravitation, where a body is surrounded by a field that warps space and time. Later, Einstein and others attempted to construct a unified field theory in which electromagnetism and gravity emerged as different aspects of the field. Unfortunately, they failed, and to this day, gravity resists any attempt to include it among the field theories of the standard model (C. Sutton (2020)).

At subatomic distances, the rules of quantum mechanics apply to the fields. For example, charged particles interact in quantum electrodynamics by emitting and absorbing photons (minute packets of electromagnetic radiation). So successful was the theory of quantum electrodynamics it became the prototype for the theories of the other forces.

According to the standard model, the basic building blocks of Nature are the "quarks" and the "leptons." Leptons, such as the electron, are free particles, not bound by external forces. On the other hand, quarks are always bound together by the strong force, which overwhelms the electromagnetic forces at subnuclear distances but is extremely short-ranged, about $10^{-15}$ $m$. The leptons are not subject to the strong force. However, quarks and leptons experience a second nuclear force, the weak force, responsible for certain types of radioactive decay but feeble compared to the electromagnetic force (C. Sutton (2020)).

As the picture of the subatomic world crystallized, theorists realized that "symmetry" was a key feature of Nature. Particular symmetries determined which particle interactions could and could not occur. Theories that include symmetries are called "gauge theories," where the equations expressing the theory's contents give the same results amid changes to the variables in the equations. Such equations are called "gauge invariant," which means that the equations are constrained by the symmetries that govern them. Both electromagnetism and general relativity are gauge theories. However, the weak force could be made gauge-invariant only if the electromagnetic force was included. Sheldon Glashow, Abdus Salam, and Steven Weinberg independently proposed a unified "electroweak" theory of these forces based on the exchange of four particles: The photon for electromagnetic interactions, two charged W particles, and a neutral Z particle for weak interactions (C. Sutton (2020)). All the theories of the standard model are gauge theories.

During the 1970s, a quantum field theory (QFT) for the strong force was developed and incorporated into the standard model. In this theory, quarks interact by exchanging particles called "gluons." Called "quantum chromodynamics," the theory has yet to be unified with the electroweak theory. Nevertheless, several potential theories, called grand unified theories" (GUTS), portend to unite the strong and electroweak forces. But, GUTS are difficult to evaluate since the energy scales at which they tell at least part of their stories are out of the reach of current or future collider experiments. Nevertheless, there is evidence that the strengths of the different forces converge at high energies.

Most GUTS predict that quarks can decay into leptons, specifically proton decay. But if the proton does decay into lighter particles, its expected lifetime is about 1032 years. The hypothesis that protons decay has been tested by monitoring large amounts of matter containing on the order of 1032 protons. Still, no evidence of proton decay has been found yet. Moreover, the strengths of the forces do not converge exactly unless the GUT includes "supersymmetry," a hypothesized symmetry between the force-carrying and the matter particles. Even so, any successful GUT will not include gravity since theorists have yet to formulate a QFT based on the exchange of the hypothesized force-carrying "graviton" (C. Sutton (2020)).

— *The Heart of Man*

# Chapter 1: The Status of Unification

***"Today scientists describe the universe in terms of two basic partial theories—the general theory of relativity and quantum mechanics. They are the great intellectual achievements of the first half of this century. The general theory of relativity describes the force of gravity and the large-scale structure of the universe, that is, the structure on scales from only a few miles to as large as a million million million million (1 with twenty-four zeros after it) miles, the size of the observable universe. Quantum mechanics, on the other hand, deals with phenomena on extremely small scales, such as a millionth of a millionth of an inch. Unfortunately, however, these two theories are known to be inconsistent with each other—they cannot both be correct."***

***— Stephen W. Hawking***

### 1.0 Introduction

Currently, the theory of relativity and the quantum theory are the two grand theories describing the nature of our Universe. But unfortunately, the theories are incompatible; both cannot be correct. Choosing between them should boil down to uncovering which one is correct. But it is hardly that simple since both theories make spectacularly successful predictions in their respective domains - relativity on the large scale and quantum mechanics on the small scale. Therefore, a theory that replaces these theories must make as many or more accurate predictions as the combined predictions of its predecessors. The current approach to solving this problem involves unifying, in some manner, the two grand theories into one. The name for this, yet-discovered, unified theory is "quantum gravity."

A successful theory of quantum gravity would require that the incompatibilities between the theory of relativity and quantum mechanics be reconciled in some fashion. Unfortunately, all current reconciliation projects expose a myriad of difficult-to-resolve issues, some of which reach beyond the domain of traditional physics. Having access to an abundance of empirical data, through pattern recognition, physicists have created fruitful conjectures as a basis for a theory. Comparing the conjectures to the physical evidence gauges how well the theoretical suppositions fit the observed facts. But pursuing a quantum theory of gravity carries physicists into territories where little or no data is available, the absence of which provides scantily few methods of verifying the correctness of a theory. And if not on verifiable evidence, what merits accepting one case for unification over another (Callender (2001) pp. 1-29)?

While the lack of data has practical implications, more problematic are the conceptual issues. For instance, reconciling the quantum concept of "quantizing" fields with the relativistic idea of "general covariance" has found no generally accepted solution. In general relativity, space-time is the entire substance, described as a dynamically smooth fabric; it can be molded, folded, and stretched but not torn. The quantization of space-time involves turning the space-time smoothness into discrete quantities; in simpler terms, turning space-time into particles. However, quantum field theory (QFT) assigns a different field to each particle, and those fields differ entirely from the gravitational field. Consequently, quantum particles propagate in a fixed space-time and are not extensions of the dynamical fabric of space-time.

On the other hand, the equivalence of gravitational and inertial mass is fundamental to general relativity. It makes gravity a property of space-time itself rather than a field propagating within space-time. Quantizing gravity subjects some of the properties of space-time to quantum fluctuations, which is the same thing as saying that space-time oscillates (Weinstein (2021), § 2). The best way to describe the quantization of fields is by specifying a specific fixed space-time background (one that does not oscillate) and then quantizing fields against that background. Such an approach makes it easier to formulate the concepts related to the fields such as energy, momentum, and angular momentum, etc. (Ashtekar (2005), § 1.2). But the space-time metric is dynamic, determines the space-time geometry, and dictates the propagation of all physical fields (including itself).

Because there is no background geometry, the concepts of energy and momentum are challenging to define. It's a "chicken and egg" problem. Since there is no fixed space-time, defining basic notions, such as causality, time, and evolution, becomes problematic without first specifying a space-time geometry (Ashtekar (2005), § 1.2). According to general relativity, the laws of physics remain unchanged with a change in space-time geometry, which is the essence behind the idea of "general covariance." But, quantum mechanically, a change in space-time geometry introduces a new set of physical laws since each particle requires its particular field. And it is hard to see how these two diametrically opposed field concepts can be reconciled.

Additionally, the conception of "time" in relativity and the quantum theory differs significantly. Quantum mechanics treats time as a parameter. All observations happen at a particular moment. Potential observables evolve in time, but the time evolution itself is not observable. The mathematics of quantum mechanics describes the evolution of a system by a "wave function." But the evolution of the wave function is not observable. Upon a measurement or observation, the wave function ceases to be a factor, and only a real outcome is observed. By contrast, general relativity's "time" must be observable since it describes, in part, the warping of space-time. In this case, time is a dimension given equal weight to the other three spatial dimensions (Callender (2001) p. 22).

There are other conceptual issues. For example, quantum theories are partly indeterministic, while general relativity is entirely deterministic. This fact and the previously discussed issues make formulating a quantum theory of gravity difficult.

Nevertheless, attempts have been made to rectify the incompatibilities. In a nutshell, the theory of relativity relies on a four-dimensional space-time. On the other hand, the theory of "weak interactions" depends on a two-dimensional spinor space. In contrast, the theory of "strong interactions" is described by a three-dimensional color space. From a unification standpoint, it is difficult to see how these three theories can be combined into one, given that the number of dimensions in each differs.

Yang and Mills (1954) presented a most promising approach involving a "gauge connection." They, among others, fostered the idea of combining seemingly disparate spaces into a new space. The formulation typically involves creating a "base space," *M* (usually space-time in most applications), then bundling the base space together with another space, *V* (in current terminology, this space is called a "fiber") to form a new space, ($M \times V$), called the "total space." The best way of thinking about this is to imagine that there is a copy of *V* (the fiber) that sits above each point of *M* (base space). The advantage of bundling is that the total space can exhibit different characteristics than either *M* or *V* considered independently. Bundling is accomplished in innumerable ways, which leads to many theories. Yang and Mills, among others, recognized that requiring the fiber to possess a specific symmetry gave the bundled spaces unique properties, reducing the number of possible theories (Penrose (2005), § 15.2).

Bundling is at the heart of the mathematics (QFT) that supports the standard model of particle physics. Nevertheless, the Yang-Mills theory was not an immediate success since the particles described by the theory were all massless, and some elementary particles were known to carry mass. The massless particle requirement resulted from the symmetries inherent in the mathematics that made the bundling technique viable. Also, as initially conceived, the Yang-Mills theory was classical. But in 1964, Peter Higgs proposed a mechanism that described how the symmetries in electroweak interactions were broken. His insights explained the origin of the masses of elementary particles in general and of the "W" and "Z" bosons in particular. The "Higgs mechanism," which had several inventors besides Higgs, predicted the existence of a new particle called the "Higgs boson" (Rincon, (2008)).

Solving the mass problem breathed new life into Yang-Mills' theory. But there was one additional hurdle. When some numbers were totaled, the sums diverged, meaning that the summations added to infinity. Mathematical techniques explaining the laws of Nature should result in finite answers. Calculations that result in infinities are disastrous. However, techniques were developed where, if certain conditions existed, the infinities could be removed or ignored by adjusting the theory's parameters. The process of removing or ignoring infinities is called "renormalization." In 1971, Gerard 't Hooft demonstrated that theories with a "spontaneously broken" symmetry were renormalizable (Penrose (2005), § 25.5, 26.9).

The enthusiasm for the Yang-Mills approach led to attempts at constructing a theory of quantum gravity using the bundling techniques so successful at the quantum level. Unfortunately, all candidate theories turned out non-renormalizable. The approach required that space-time be broken into two spaces; one space represented a flat space-time (the base space), and the fibre space was constructed to employ the successful perturbation techniques associated with the quantum theory. But introducing a background metric was incompatible with "general covariance," which is fundamental to general relativity.

To appreciate the significance of this, consider the electroweak theory. This theory is perturbatively renormalizable. Although individual terms in the perturbation expansion (summations) may diverge, the infinities can be systematically absorbed in the values of the free parameters of the theory. By renormalizing these parameters, individual terms in the perturbation series can be systematically rendered finite. In general relativity, such a systematic procedure fails. The infinities that arise are genuinely troublesome. Put differently, a quantum theory of gravity approached this way acquires several undetermined parameters that cannot be absorbed (Ashtekar (2005), § 1.1).

Further attempts were made, specifically a theory called "$N = 8$ supergravity." For several years, supergravity fostered confidence that it was on the threshold of providing a complete quantum gravity theory. The theory displayed a less volatile ultraviolet behavior. But ultimately, it, too, was non-renormalizable (Ashtekar (2005), § 1.1).

Other approaches included affixing additional dimensions to a given space (usually space-time). Theodor Kaluza (1921) extended general relativity to a five rather than a four-dimensional space-time. His equations were separated into further sets of equations, one equivalent to Einstein's gravitational field equations. The other set was equivalent to Maxwell's equations for the electromagnetic field. The theory could not be successfully extended into the quantum realm, but the idea of extra dimensions persevered.

Another idea was to replace the point particles of the standard model with 1-dimensional objects called "strings" and associate particle-like states with the string's various "modes of excitation." The approach has been the dominant candidate for creating a quantum theory of gravity for the last twenty or thirty years. It includes the spin-1 modes characteristic of the gauge theories associated with atomic interactions and a spin-2 massless excitation, which automatically incorporates the graviton (the fundamental particle thought to carry the gravitational force). In this sense, gravity was already built into the theory! Moreover, Greene and Schwarz (1987) showed that perturbative string theory was consistent in a certain number of space-time dimensions – 26 for a purely bosonic string and 10 for a superstring (Polchinski (1998)).

Strings live in a flat space-time background, where perturbative estimation techniques apply. However, since it includes extended objects rather than point particles, all singularities (the cause of the infinite sums) are eliminated, dramatically improving the theory's ultraviolet behavior. It is widely believed that the perturbation techniques applied to the theory are all finite. In short, the theory has all the ingredients of a "***theory of everything***" (Ashtekar (2005), § 1.1).

However, string theory faces several challenges: (i) To date, none of the string theories created so far models the observed world; (ii) it is not certain that the perturbative expansions associated with string theory are, in fact, finite. For example, Gross and Periwal (1998) have shown that in the case of bosonic strings, when summed, the series diverges (sums to infinity) and does so uncontrollably. Moreover, they also argued that the conclusion would not change in the superstring theories.

Why is the divergence of a sum regarded as a serious failure of string theory but accepted in other quantum theories, such as quantum electrodynamics (QED)? QED is not a complete theory and has the advantage of ignoring many processes that come into play at high energies. For example, it ignores the microstructure of space-time by assuming that a smooth continuum below the Planck scale can approximate space-time. On the other hand, a "***theory of everything***" is not at liberty to ignore anything and must resolve the Planck problem non-perturbatively. Moreover, experiments have not yet verified that the extra dimensions associated with supersymmetry are properties of Nature. Nonetheless, string theory provides a glimpse of an entirely new vista, the concrete possibility that unification results from a tightly woven, non-local theory (Ashtekar (2005), § 1.1).

## 1.1 The Search for a Unified Theory

Modern physics dates from the 17th century, beginning with Isaac Newton's theory of gravity, followed by Maxwell's theory and Faraday's description of electromagnetism. Extensions and refinements to the work of Newton and Maxwell resulted in non-relativistic classical physics, which succeeded in unifying most of the physics we experience daily. This unification explained how things move, to jet and rocket propulsion, stresses on bridges, buildings, and towers, electricity, sound, light, and many other phenomena too numerous to mention.

Simple Natural laws expressed mathematically describe classical physics. The "particle" represents one of the two significant modes of describing physical phenomena; the other mode is the "wave" or "vibration." Rigid massive bodies are generally described as particle-like, while light, for example, is described as a wave or, more generally, by the "field" (Halliday (1988), § 17-1).

Maxwell's theory united all previously unrelated observations, experiments, and equations of electricity, magnetism, and optics into a consistent theory of electromagnetism. Moreover, Maxwell's laws, like Newton's, expressed almost entirely in mathematical language, described electromagnetic phenomena as wave-like (field) (Nahin (1992), p. 45).

The Twentieth century saw the emergence of the theory of relativity (our current theory of gravity) and quantum mechanics. In the process, two new forces (the strong and weak nuclear forces) were discovered. However, the era's crowning achievement was the "standard model of particle physics," which combined electromagnetism and the two nuclear forces into one model. If physics is to be unified into one theory, the last remaining step is to merge gravity, the fourth of the four fundamental forces, with the standard model.

Optimism grew that a unified theory might emerge from string theory. In the search for a successful unified theory, string theory is considered the most likely candidate. According to the standard model of particle physics, fundamental particles—electrons, for example, are "point particles" which do not take up space. In this sense, physical space is regarded as infinitely divisible, where any region in space can be further subdivided, no matter how small.

However, at a specific point, the equations of general relativity produce infinities that cannot be ignored. String theory overcomes the "infinity problem" by proclaiming that matter consists not of point particles but tiny loops of string. In this sense, strings make space-time fuzzy, removing the need to explain what happens at a specific point. From this beginning, the laws of physics emerge. General relativity, electromagnetism, and Yang-Mills gauge theories appear surprisingly. But string theory comes with other ingredients, most strikingly, extra spatial dimensions beyond the three observed (Tong (2009), p. 1). Moreover, instead of one theory, there seems to be a vast landscape of possible theories – perhaps $10^{500}$ or more.

The emergence of string theory has fractured the physics community into several ideological factions. Some physicists claim that string theory has failed, makes no testable predictions, and is unlikely to do so. They criticize the theory as "not science," arguing that the vast landscape of theories makes it impossible to evaluate the theory (Woit (2006), § 12, Smolin (2007), § 10). Conversely, some physicists have adopted a "wait and see" attitude, openly admitting that the evidence for string theory is circumstantial. Still, the hope is that the circumstantial evidence will one day bear the fruits of its potential (Smolin (2007), § 10). And finally, a different set of physicists embrace the idea of a landscape of theories.

Many physicists who endorse the idea of a landscape argue that the theory that describes "our Universe" will be found by applying the anthropic principle (Susskind (2003), § 1). Roughly speaking, the anthropic principle subscribes to the notion that there are finely-tuned connections between the physical parameters of a theory and the conditions necessary for life to exist. In other words, the input parameters to the theory are confined to a narrow range of values. As the anthropic argument goes, these connections allow choosing among potential string theories only those with the parameter values that allow life to exist (Jones (2010), § 11).

### 1.1.1 A Rigorous Approach to Sausage Making

Theoretical physics entails writing down several Natural laws expressed through a set of mathematical equations, then arguing that the equations faithfully represent the observable facts. Repeatedly observing the same or nearly identical results under similar observational or experimental conditions implies that Natural laws are eternal truths. Several Natural laws have yet to be violated in any previous experiment, Coulomb's law, for example. Still, there is no guarantee that, in the future, a Natural law will not be subject to modification. This has been, roughly speaking, the general approach to physical science. Natural laws are the most probable (Halliday (1988), § 22-12).

Science, approached in this manner, more or less embodies "falsification," the idea that if Natural laws, expressed by the equations of a theory, faithfully represent physical phenomena, then those equations should make testable predictions. Observations and experiments verify that the predictions of a theory either do or do not align with those observations. Traditionally, if there is no way of empirically testing a theory, it is generally considered a conjecture. The upshot is that a theory can never be proven but should be capable of being disproven.

To a certain extent, the concept of a "Natural law" is confounded by the uncertainty principle, the idea that specific system properties can never be known with certainty. Conservation laws constitute the underlying fundamentals of physics. Ironically, conservation laws are routinely violated, but those violations occur within the dictates of the uncertainty principle. Poignantly, the uncertainty principle acts as a supreme Natural law. All other Natural laws are subjected to it. Put differently, no Natural law can be written more exactingly than the uncertainty principle permits.

Such an outlook suggests physicists work in cookbook environments where rigorous methods produce irreproachable results. But physicists are prone to human frailties and prejudices like the rest of us. The history of physics is replete with instances where debates can devolve into bitter disputes, mainly if there is scant evidence supporting an argument. Personalities, individual prejudices, and personal clashes have often led to mistakes and deadends and created impediments to new discoveries. Obstinately refusing to accept new but ultimately correct views has resulted in embarrassing personal confrontations, sometimes contributing to the slow advancement of the subject. Yet, through all the sausage-making of theoretical physics, the end product reflects something reasonably close to the truth.

### 1.1.2 A "Theory of Everything?"

Unification means finding a single set of consistent and straightforward Natural laws that describe our Universe. In just over a century, remarkable progress has been achieved in the face of enormous challenges. Advancements in sophisticated technologies make collecting mounds of critical, if disparate, data possible. Analyzing the data has led to discoveries and generated the enthusiasm necessary to uncover the mysteries lurking in the data. For example, a century ago, discussions and prognostications were limited to explaining how our solar system works. Today, theories sporting complicated, sometimes esoteric ideas, foretell how the entire Universe works.

Some physicists believe that a ***"theory of everything"*** is at hand. Unifying the strong, weak, electromagnetic forces and gravity has been at the forefront of physics for about four decades. But, of course, the "***theory of everything***" is not remotely a theory of everything. The term is primarily an invention of the Press. The term "***theory of everything***" is a euphemism for the single theoretical framework that would explain all the fundamental interactions of Nature (Gross (2005)).

### 1.1.3 Physics or Speculation?

The desire to unify physics gave birth to several theories sporting various speculative ideas. While interesting, the new theories arrive with a host of mysterious ingredients (Tong (2009), § 0), are often highly abstract and frequently accompanied by meritorious declarations, claims that are often difficult to substantiate, outright dubious, or, in some cases, just plain ridiculous. For example, the currently proposed solutions to unifying the quantum theory with the theory of relativity describe physics at energy levels that are experimentally out of reach, even if future technological developments materialize. Some physicists find it unprofessional to prognosticate in realms where experimental observations are sidelined, replaced by conjectures whose advocates point to esthetic beauty as proof that the theory is correct.

To be taken seriously, a theory should make predictions that are, at least potentially, falsifiable. Interestingly, in pursuing the ultimate theory, the "strictly falsifiable" criterion has increasingly been ignored. Ideally, a good theory should be amenable to experimental verification and provide new insights into how Nature works (Smolin (2007), § 2). Relatively strict adherence to such tenants ensures that prognostications tread on comparatively safe ground. Ideally, there is less risk of error, and theoretical predictions are based on something more than philosophical speculation.

But much of the current theoretical work bears little resemblance to the characteristics just mentioned. Many conjectures today are often unfalsifiable and make no truly insightful predictions or, in some cases, no predictions. Whether the "strictly falsifiable" criterion should be levied on current theoretical work is debatable. But lacking sufficient substance, how should the merits of various unifying proposals be evaluated?

Lee Smolin, the author of "*The Trouble with Physics*," is greatly perplexed by how his profession currently supports physics research projects. Smolin argues that the current funding methods demote the pursuit of ideas to an exercise in autocracy. He contends that the funding methods have split the physics community into factions of similarly thinking groups. Part of a physicist's job, as a group member, is to attract other talented thinkers to join. More group members equate to a higher number of published research papers that reflect the commonly held beliefs of the group. Naturally, the popular idea rapidly advances at the expense of ideas that lack sufficient representation (Smolin (2007), § 19).

For many physicists, including Smolin, the environment is distasteful. From their perspective, the danger is that theoretical physics becomes guided by authority rather than by experiment and observation. For example, the theoretical physics departments at most major universities are dominated by string theory supporters, which, it is argued, creates a "string theory cult" that inherently marginalizes other approaches. Science, like any other field, needs criticism. Psychologists have shown that "groupthink" takes hold where the only people allowed a seat at the table are those who think alike. To have a robust intellectual exchange, others that challenge the popular viewpoint need inclusion (Jones (2010), p. 308).

The primary method for determining a proposal's merit is peer review; this, as mentioned above, comes with a unique set of issues. Modern theories addressing unification are highly technical, some exclusively so. It is difficult, in many cases, to find experts with the requisite knowledge to judge the merits of an idea adequately. Consequently, physicists often complain that many ideas judged worthy by peer reviewers merit no such worthiness. On the other hand, some ideas of merit are rejected or ignored.

The problem is exacerbated because modern theories in physics are temples of extreme abstraction, understood by few individuals. Gone are the days when theories involved forces and masses, rods and pulleys. A baffling array of mathematical structures replaces concrete mechanical concepts. In a certain sense, modern particle physics is entirely incomprehensible. The old physics with points, pushes, and pulls is there somewhere if only hidden behind the mathematical concepts of fields, interactions, wave functions, and symmetry groups. The modern theoretical physicist has much in common with the post-modern deconstructionist. Both wallow in their special gobbledygook, understandable to the specially trained but inaccessible to the uninitiated. To many outsiders, modern theoretical physics can seem like an impulsive subject where aesthetic beauty trumps scientific fact, and increasing abstraction turns theoretical physics into recreational mathematics (Lindley (1993) pp. 18-19).

The current environment is not unlike the epic periods of past generations, where similar doubts have stunted progress. Even Maxwell's theory was slow to be accepted. He tried postulating that little mechanical wheels were responsible for the swirling effect of electromagnetic phenomena. But, in the final publication of his work, the wheels vanished, replaced only by the concept of the "electromagnetic field" and leaving no explanation as to what the source of the field was or how the force field moved through space (Lindley (1993) pp. 37-43). Traditional thinkers tried retaining the field's mechanical nature by postulating that it moved through space in a "luminous aether." They clung to the idea that light waves, like sound or water waves, required a substance through which to travel. When carefully designed experiments failed to find the aether, it was the first step away from a Universe described mechanically and one step toward a Universe explained through a heightened level of mathematical abstraction.

Current unification proposals have become so broad that the only feasible experimental laboratory available, realistically speaking, is the entire Universe, placing ever-increasing importance upon the science of "cosmology." Cosmologists seek to understand the origin, evolution, structure, and ultimate fate of the Universe at large, as well as the natural laws that keep it in order. Modern cosmology is dominated by the "big bang theory," which combines observational astronomy and particle physics (Battersby (2006)).

Of all the physics-related disciplines, cosmology is the newest and most tenuous. Given the enormous distances from the Earth of most celestial objects, it is exceedingly difficult to estimate how far away, how bright, or even to identify what the objects are. As recently as the 1920s, astronomers were unsure whether the Universe was much larger or confined to the Milky Way galaxy. Today, it is known to be much larger, but how much larger remains an open question. The current practice of astronomy requires exacting technology coupled with ingenious methods of calculation, where gross statistical estimates are typical. The discoveries of the optical and radio telescopes and the photoelectric plate dramatically enriched the science. The photoelectric plate replaced the unreliable human eyeball, greatly enhancing the consistency of celestial observations. But comparatively speaking, cosmological theories are the most precarious, even aided by the astronomer's propensity for exacting detail. Astronomical Estimations of the age of the Universe have been modified (by billions of years) at least three times within the last fifty or so years, primarily due to better methods of estimating celestial distances (Singh (2004), § 5). The slightest astronomical insight can dramatically affect how the Universe is perceived.

There is no agreed-upon list of criteria or even general guidelines on how potential unification theories should be evaluated. But there are boundaries and constraints to which potential solutions must adhere. Technological limitations constrain the extent to which laboratory experiments can act as judge and jury over competing theories. While helpful, there is little hope that new technologies can penetrate the energy scales where these theories tell at least part of their stories. For all intent and purposes, expecting that modern unification theories be entirely or even primarily falsifiable is unreasonable. On the other hand, modern physics has uncovered fundamental laws covering a wide range of physical phenomena through the quantum theory of matter, Einstein's theory of space-time, and the big bang theory of cosmology. It is imperative that the understanding gained from these theories be incorporated into any potential unifying proposal.

There are other significant hurdles to overcome. Beyond reconciling the general theory of relativity with quantum mechanics, the objects in the Universe are accelerating away independently of vantage point, presumably in a manner consistent with the second law of thermodynamics. In a nutshell, the second law says that the Universe evolves from a state of order to a state of disorder as time advances. This process, called "entropy," is irreversible. Perhaps no law of Nature commands greater credibility than the second law, summed up in the following quotation:

> *"The law that entropy always increases holds, I think, the supreme position among the laws of Nature. If someone points out to you that your pet theory of the Universe is in disagreement with Maxwell's equations – then so much the worse for Maxwell's equations. If it is found to be contradicted by observation – well, experimentalists do bungle things sometimes. But if your theory is found to be against the second law of thermodynamics, I can give you no hope; there is nothing for it but to collapse in deepest humiliation."*

- Sir Arthur Stanley Eddington

Imagine rolling back the clock to the time of the big bang. According to the second law of thermodynamics, the Universe must have been incredibly orderly at that moment. So orderly that any model describing the "real" Universe would require input parameters so exacting that the tiniest variation in those initial conditions would create a model universe vastly different from the one observed.

The challenge comes in describing the role of quantum mechanics in the early Universe. The fundamental principles of quantum mechanics don't allow the clock to be turned back in the way suggested by the second law. An unresolved problem in physics, often called the "vacuum catastrophe," is that most quantum field theories predict a massive value for the quantum vacuum. It is reasonable to think that the big bang is somehow related to this quantum vacuum state. A common assumption is that the quantum vacuum is equivalent to the cosmological constant, which is fundamental to modern cosmology (Rugh (2001), pp. 663-701). But, if an effective local QFT describes the Universe down to the Planck scale, the theoretical value of the cosmological constant is 120 orders of magnitude greater than its measured value. This discrepancy has been called "the worst theoretical prediction in the history of physics!" Worse yet, there is no known way of deriving a tiny cosmological constant from particle physics (Hobson (2006). p. 187).

Moreover, observations have shown that objects in the Universe, such as stars and galaxies, are accelerating away from all observation points. This phenomenon is happening in a Universe where gravity is the dominant force. Common sense would dictate that gravity should have the opposite effect. How can the Universe expand if the physics at these large scales is dominated by gravitational forces, which should counter such an expansion? The general theory of relativity does not require the Universe to be static. Still, it leaves an open-ended description regarding the dynamics of the Universe. It is unclear whether the Universe will expand forever, eventually contract into a big crunch or remain pretty much as it is. Any serious unification theory should reasonably explain this expansion/gravitational effect.

Finally, there are many unsolved problems: Finding an explanation for dark matter/energy, the black hole information paradox, supersymmetry, and many others. Any unification proposal will be judged partly on how it addresses these issues.

At this point, the recipe for creating an adequate unified theory of physics will involve some confirmation by observation and experiment, consist of clever reasoning, and be acceptable to peers. Proposals will certainly arrive laced with healthy doses of speculation. Hopefully, methods will arise for determining whether or not the proposals are acceptable, other than on purely authoritarian grounds.

With any luck, the final theory, if one emerges, will become something more than a myth. A myth is a self-contained belief that explains everything but can be neither proven nor disproven. And therein lays the danger. A myth is something everyone agrees upon because it is convenient, not because it is true (Lindley (1993) p. 255).

## 1.2 Destiny: Will There Be Unification or Not?

At a crossroads, the quest for unification appears destined for one of three possible futures:

1. The hope for Unification is abandoned, replaced by adopting several theories.
2. Current approaches will eventually lead to a satisfactory unified theory (e.g., string theory).
3. A satisfactory unified theory requires an entirely new approach, which is currently unknown.

### 1.2.1 First Future - Plurality: The Death of Reductionism

Those who pursue a unified theory are often called "reductionists." Like the monist, a reductionist subscribes to the philosophical view that reality is grounded in a single primary substance or principle. Accordingly, only one theory should explain how that "one substance" works. Currently, two incompatible theories explain how the Universe works, and this, to a reductionist, is intolerable.

But to someone who does not subscribe to the reductionist view, the existence of two incompatible theories is not at all disingenuous. Non-reductionists often argue that the essential role played by higher organizing principles in determining emergent physics is generally overlooked (Laughlin (2000)). Furthermore, any single unifying theory is unlikely to describe Nature completely. Those who think along these lines find no compelling reason why the laws of Nature should require the same structure and do not share the view that all phenomena emerge from a single underlying principle.

Non-reductionists argue that the essential role played by higher organizing principles in determining emergent behavior "continues to be disavowed by so many physical scientists" that the situation equates to a poignant comment on the nature of modern science. They point to unpredictable electronic phenomena such as organogels, Kondo insulators, and cuprate superconductivity as prominent examples, "not discussed in polite company," because these ideas are contrary to the reductionist beliefs so central to physics (Laughlin (2000)). They accuse their colleagues of acknowledging only the facts supporting the reductionist case, arguing that this is fundamentally incompatible with how science should be conducted.

Non-reductionist sentiments can be summed up in the following quotation:

> *"Cosmologists think that cold dark matter is a safe bet because particle physicists can provide it; particle physicists invent theories containing cold dark matter because cosmologists seem to want it. Cosmologists think the Universe must have exactly the critical density, despite observational evidence, because inflation demands it; particle physicists think inflation is a marvelous idea because it predicts a Universe of precisely the critical density, which is what the cosmologists want. The hopes of the cosmologists and the particle physicists have become the same hopes, and their methods have become the same method: on both sides of the joint effort there is absolute reliance on the notion that a single theory will explain everything, and when such a theory comes along it will be instantly recognizable by all. This might be called the messianic movement in fundamental science."*
>
> - David Lindley

In addition to those physicists who reject the reductionist view, the "instrumentalists" adopt an agnostic attitude toward reductionism. From their perspective, physical theories should primarily make credible predictions. Theories should not be expected to uncover a reality in any more profound sense. Those "deeper" questions should be left to the philosophers and the churches (Callender (2001), pp. 3-

4). Since the theory of relativity and quantum mechanics make accurate predictions in their realms of applicability, instrumentalists see no reason to pursue a unified theory and regard such an endeavor as needless speculation.

But the quantum theory predicts that matter arises from matter fields. The theory of relativity reveals that matter causes space-time curvature and, hence, matter should create a space-time curvature at high energies. Neither the theory of relativity nor the quantum theory adequately addresses this situation. Even an instrumentalist would not likely argue that the two theories should not be modified, at least to the extent that the correct physics is predicted (Callender (2001), pp. 3-4).

### 1.2.2 Second Future: A Unified Field Theory

Despite its predictive successes, the standard model of particle physics is not the final answer. The situation has motivated a search for a higher level of unification. An enhanced standard model should:

4. Include gravity and explain dark energy/matter.
5. Show why the 19+ free parameters have the values they do.
6. Predict the correct masses of the neutrinos.
7. Develop a genuinely finite theory.
8. Show why there are 3 generations of leptons and quarks.

That the standard model does not include gravity has been a known shortcoming since its inception. But beyond that, the model requires 19 or so adjustable constants. The laws of the theory require specifying the constants. The values of the constants are found by experiment. But there is no known explanation for why the constants have the values they do. Every time the model is run, the values of the constants must be specified (Smolin (2007), pp. 12-16). Each adjustment represents a basic physical fact that the current theory does not explain.

The theory of relativity and the quantum theory suffer from the "problem of infinities." To many sensible questions put to these theories, the answer often comes back "infinity," while observations or experiments suggest otherwise. The situation implies that both theories contain fundamental flaws or at least significant shortcomings. The standard model handles the "problem of infinities" through "renormalization." Renormalization involves methods where infinities are canceled or ignored.

Understanding renormalization will require a bit of a detour. First, what is it that is renormalized? The answer is the "Lagrangian." More to the point, the parameters of the Lagrangian are renormalized. The Lagrangian summarizes the dynamics of a system and is defined as the kinetic energy, $T$, of a system minus its potential energy, $V$. In symbols, $L = T - V$.

Finding a serviceable Lagrangian is crucially important because the energy of a system can be represented without requiring the system's status to be known at every moment. Classical physics demands that the status of a system be known at every moment (position and momentum). But quantum processes are uncontrollable. It is impossible to know the system's complete status at each moment. Therefore, classical equations that require this cannot be employed at the quantum level (Bohm (1951), § 23).

Although the Lagrangian was initially developed to describe classical mechanics, it applies to quantum mechanics (more on this later). If the Lagrangian of a system is known, then the system's equations of motion can be obtained.

To gain a sense of renormalization, consider the following problem: in physics, the notion of a "velocity" is described in terms of distance and time (i.e., $s\ (distance) = v\ (velocity) \times t\ (time)$). Suppose I am driving a car and want to know how fast I'm going right now. How would a physicist answer this question? If it is "right now," then no time ticks off the clock ($t = 0$), and, in that time, I go nowhere, so $s = 0$. Hence, $v = 0/0$. But my speedometer reads, say, $69.3$ *miles per hour* right now, so the physicists' answer of $v = 0/0$ is nonsense.

How does a physicist correct this obvious incorrect result? Instead of this very instant, suppose I measure how far I travel in, say, $t = 0.001\ hr$. Further, suppose in $t = 0.001\ hr.$, I travel $0.0692\ miles$ (i.e., $s = .0692$ miles), then $v = 69.2\ (0.0692/0.001)\ miles\ per\ hour$.

This result is a better answer, but still not exact. Of course, $t = 0.001\ hr.$ is not instantaneous, so to get a better estimate, I could measure the distance traveled in $t = 0.0001\ hr$. This time interval is closer to instantaneous than $t = 0.001$, and if the distance traveled in that time is $s = 0.006927\ miles$, then

$$v = 0.006927/0.0001 = 69.27\ miles\ per\ hr.$$

The time, $t = 0.0001\ hr$ is closer to, but still, not instantaneous. I could repeat this process by choosing $t = 0.00001\ hr.$, and, if in that time, $s = .0006929\ miles$, this implies that $v = 69.29\ miles\ per\ hr$.

Notice as the time intervals get shorter, my speed approaches $69.3\ miles\ per\ hr$, my actual speed. So I conclude that my actual speed is $69.3\ miles\ per\ hr$. This is an assumption on my part because I didn't measure $v$ at $t = 0$. Instead, I extrapolated out from the measurements I had taken previously. Theories amenable to this kind of procedure are called "renormalizable." While renormalization procedures are generally more complicated, this somewhat pedestrian example gives a sense of what the renormalization process entails (Metzler (2011)).

Renormalization has become a necessary feature of quantum field theory because observable quantities in quantum theories, $E(x,t)$, for example, which designates the electric field strength, fluctuate from measurement to measurement and from point to point. An exact value for $E(x,t)$ cannot be obtained. The best that can be done is to compute the average value of $E(x,t)$, say, within the area of a circle (or sphere) of radius, $a$. Mathematically this can be represented by

$$\left\langle 0 \left| \left(E(x+a,t) - E(x,t)\right)^2 \right| 0 \right\rangle \approx \frac{1}{a^4}.$$

As $a \to 0, 1/a^4 \to \infty$. The quantum electric field cannot be defined under these circumstances. For example, derivatives of the field, $E(x,t)$, cannot be defined where the difference, $E(x + a) - E(x)$, diverges as the separation ($a$) vanishes. This is a fundamental problem and a consequence of the uncertainty principle (Lepage (2005), p. 2).

Renormalization starts with the process called "regularization." It is accomplished by removing from the theory all states having energies much larger than some cutoff value, say $\varepsilon_0$. This limitation effectively removes the infinities (Lepage (2005), p. 2), which is similar to replacing $t = 0$ with $t = 0.001$ in the "car velocity" example. Making that replacement got rid of the nonsensical answer.

However, regularization comes at a price. There is no way of guaranteeing that cutting off part of the short-distance scale will have little or no impact on the long-distance behavior of the theory. If the short distance were relevant to long distances calculations in any given problem, it would introduce and often does introduce unacceptable errors in the calculations.

The basic idea behind renormalization is that all effects of the very high-energy states on the low-energy behavior of the theory can be discarded, provided the theory's Lagrangian is modified to account for the effects that result from the discarded states. To accomplish this, a Lagrangian ($L_0$) density function is defined, which describes the energy states of the system together with a regulator, $\varepsilon_0$ (cutoff), that truncates the theory's state space at some very large value. The other parameters specified by the Lagrangian remain well-defined so long as $\varepsilon_0$ is kept finite. These parameters are referred to as "bare parameters." To compute the bare parameters for a particular $\varepsilon_0$, choose two convenient processes or quantities, and compute them in terms of the bare parameters using $L_0$, then adjust the bare parameters until the theory and experiment agree (Lepage (2005), p. 3). Again, this is similar to the "car velocity" example where, as shorter and shorter time intervals were chosen, the answer got closer and closer to the measured answer.

To understand the role of the cutoff in defining the theory, start with $L_0$ and a cutoff, $\varepsilon_0$. Remove from this theory all states having energies or momenta larger than some new cutoff, $\varepsilon$ ($\varepsilon \ll \varepsilon_0$). Repeat the process described above to determine if the energies discarded in the first iteration would impact the original calculations of interest in the long range. If significant errors are found, $L_0$ must be modified to account for the previously discarded energies, which produces a new Lagrangian, $L = L_0 + \delta L_0$, where $\delta L_0$ represents an additional adjustment that more closely agrees with experiment. Only a finite number of adjustments are required when a theory is renormalizable. The result is a theory without infinities (Lepage (2005), § 2.2).

The infinities that appear in the series expansion calculations associated with quantum field theories result from the mathematical approaches used in the calculations. Physics drives renormalization. Terms within the mathematical expansions are discarded or adjusted only when it makes "good physics sense" to do so. Moreover, formalized methods have been developed for determining if the Lagrangian associated with a field theory is renormalizable. In fact, renormalization has become useful in determining if new physics exists beyond the cutoff energies associated with a given theory (Zee (2010), § VI.8).

Not all theories are renormalizable. Sometimes, all manner of adjustment fails to tame the infinities (a specific quantity cannot be approached). The general theory of relativity, for example, is non-renormalizable. No known way of modifying general relativity would allow it to describe the physics at extremely short distances.

If a given theory is non-renormalizable, it indicates that beyond a cutoff value, the calculations of the theory cannot be trusted and signifies new physics past the cutoff. As it relates to unification, renormalization is crucial. For example, the fact that the theory of general relativity is non-renormalizable indicates that the physics beyond a specific cutoff is entirely different than what general relativity says it is, providing evidence that general relativity cannot describe physics at tiny scales. This is the primary reason physicists scurry to the safer harbingers of string theory.

During the latter half of the twentieth century, theorists recognized the importance of "symmetry" in understanding the Nature of things and began incorporating it into their work. A symmetry is a feature (observed or intrinsic) of a system that is preserved under some change. An important example of a symmetry is preserving physical laws under arbitrary space-time transformations in general relativity.

Noether (1918) showed that every conservation law is associated with a continuous symmetry. Because conservation laws are fundamental to theoretical physics, this was a celebrated discovery. There are many different kinds of symmetries. The standard model, for example, contains three of them denoted by, $U(1)$, $SU(2)$, $SU(3)$. The $U(1)$ symmetry is associated with electromagnetic interactions, $SU(2)$ with weak nuclear interactions, and $SU(3)$ with strong interactions.

Lagrangian formalism is tied closely to Noether's theorems. If the Lagrangian is invariant (preserved) under a given symmetry, then the accompanying equations of motion are also invariant. This symmetry is called a "gauge symmetry" and helps show that theories are consistent with either special or general relativity.

Notably, however, Nature is not symmetrical. The process by which a system, described symmetrically, ends up in an asymmetric state is called "spontaneous symmetry breaking." Quantum mechanics seems to demand that symmetry breaking take place at some point. For spontaneous symmetry breaking to occur, a system is described as having several equally likely outcomes. Therefore, the system as a whole is symmetric with respect to those outcomes. However, if the system is interacted with in any way, the result is one specific outcome. While knowing the system is symmetric, it never displays this symmetry but ends up in one specific state (Weinberg (1993), pp. 153-160).

Spontaneous symmetry breaking plays a vital role in elementary particle interactions. The standard model contains several particles. However, within the usual Yang-Mills approach, particles are massless, while, in reality, some particles in the standard model have a mass. This omission is a shortcoming of the theoretical part of the model. The idea of spontaneous symmetry breaking is used to explain how particles gain mass. It also predicted the existence of the recently detected Higgs boson (Anderson (1962), pp. 39–442, Englert (1964), pp. 321–323, Higgs (1964) pp. 508–509). The fact that Nature appears in part asymmetrical complicates the job of creating a unified theory because it levies a requirement for finding an explanation for broken symmetries.

In collaboration with the Oxford mathematician Roger Penrose (1965), Steven Hawking showed that the Universe started with a "singularity" – a dimensionless mathematical point of infinite density. Suspicions had been swirling for some time that this might be the case. Still, it was unclear whether the singularity was inevitable or a consequence of the assumptions that went into the theory. However, all doubt was removed when Penrose proved that the singularity conjecture was inescapable. Moreover, hawking's and Penrose's demonstration was completed entirely within general relativity.

This discovery had two significant consequences. Firstly, it showed that the theory of relativity alone could not provide all the answers to the question of physical unity. The idea that the Universe started at a singular point of infinite energy seemed absurd. More likely, it was general relativity that broke down. Past a cutoff, general relativity gave absurd answers that could not be trusted. Secondly, it put general relativity and quantum theory on an unavoidable collision course. The quantum theory does not allow for the exact specification of an amount of energy at a point. The uncertainty principle within quantum mechanics forbids a singularity. This put general relativity and quantum theory into direct contradiction and illustrated the importance of finding some way of reconciling them.

Physicists desire natural theories, where the macro part of the theory follows naturally from the microscopic part. Most physicists find it undesirable, indeed unlikely, that the microscopic theory contains various free parameters that must be adjusted to give the macro part of the theory its unique properties. What they want is a compelling theory without having to fine-tune the various parameters of the theory. This puzzle is called the "hierarchy problem." Unfortunately, theorists have yet to uncover a suitable theory that does not require fine-tuning, at least to some extent (Giudice (2008), pp. 155-178). It was thought that string theory or M-theory might provide an approach free of fine-tuning, but this hope has faded somewhat in the last few years.

For all intent and purposes, from the standpoint of naturalness, the standard model is unnatural. Physicists have devised mathematical inventions to adjust the model to agree with experiments. But to argue that renormalization or, for that matter, spontaneous symmetry breaking is fundamentally the way Nature works borders on the ridiculous.

For example, the Higgs mechanism is an entirely unnatural invention. The hunt for the Higgs boson is not like a neighborhood Easter egg hunt. There is really nothing to find. The Higgs particle is a boson, and bosons cannot be detected in particle accelerators. The best that can be hoped for is that by smashing particles together at nearly the speed of light, the numbers of directly detectable particles emerge in percentages predicted by the standard model. An argument can be made that since the numbers of particles that can be seen show up in percentages that agree with what the Higgs mechanism predicts, the invention must be on the right track.

The preference is for a more natural theory, where concepts like "spontaneous symmetry breaking" emerge from the constraints of the theory and are free from inventions for which there is no theoretical explanation. Nevertheless, it is a safe bet that some physicists will continue searching for a theory beyond the standard model that gives a more natural explanation, notwithstanding the success of its experimental predictions. To that end, the search continues for a unified field theory.

The overwhelming favorite, some might argue the only candidate for unification, is string theory. Thousands of researchers are working on it. It is where most physicists, at least those who think a unified theory is essential, are placing their bets.

The premise of string theory is that matter does not consist of point-like particles but tiny loops of string. The strings vibrate like the strings of a musical instrument. Each of the different harmonics of the string represents a different particle. From this beginning, the laws of physics emerge, general relativity and the Yang-Mills gauge theories that describe nuclear interactions (Tong (2009), § 0). For this reason, string theory is a potential "***theory of everything.***"

From a unification perspective, string theory offers many alluring features. All the forces and particles of Nature can be explained as characteristic vibrations and actions of the string. The different particles arise from different resonance vibrations of the string. The vibrations of open strings are consistent with the gauge principle, which is responsible for the electromagnetic and nuclear forces. It includes gravity, which is explained by the vibrations of closed strings. It unifies bosons with fermions - the two classifications of "particles/forces" found in Nature. In most theories, the laws of motion of the particles are separate from those that govern the forces acting on the particles. This situation is undoubtedly true of classical physics. But in string theory, once a description of how strings move is given, forces arise naturally by the breaking and joining of the strings. There are but two fundamental quantities: the "string tension," which describes the energy in the string, and the "string coupling constant," which gives the probability of a string breaking into two strings or two combining into one. In addition, string theory is governed by one simple principle: as a string moves through space, it draws out a two-dimensional surface in space-time. The string moves so as to minimize this surface area, constituting the whole of the Natural law governing strings (Smolin (2007), § 12).

String theory overcomes the pitfalls that plagued previous approaches to unifying quantum theory with general relativity. Particles in QFT are conceived as existing at an arbitrarily small point. But the smaller the scale, the more turbulent quantum fluctuations become. At short distances, it is impossible to reconcile general relativity's smooth and continuous space-time with the violent fluctuations of the quantum theory. String theory limits how small the particles (strings in string theory) can become. The lengths of most strings are on the order of

$10^{-33}$ *cm*, about the size of a Planck length. Limiting the size of a string tames the quantum fluctuations just enough to allow compatibility with general relativity (Greene (1999), § 6).

Furthermore, the supersymmetric string theories seem free of the mathematical anomalies plaguing previous unification attempts. Supersymmetry dramatically enhances the aesthetic beauty of a theory. Most physicists believe that Nature uses all mathematically possible symmetries because it creates a highly aesthetic and tightly woven theory. But beyond purely aesthetic considerations, there is the practical issue of explaining quantum fluctuations. At present, the parameters of the standard model must be finely tuned to about one part in $10^{15}$ to adequately deal with these fluctuations. Because each particle contributes to the quantum frenzy, fine-tuning is required to cancel these effects. However, pairing each boson with a fermion (supersymmetry) dramatically enhances the canceling effect. It gives a much more natural explanation for what is experienced in Nature and removes the necessity for fine-tuning the standard model parameters.

In addition, at high energies, the strengths of the three forces of the standard model are almost, but not quite, equal. When supersymmetry is included, the discrepancy in their strengths vanishes due to the quantum effects of the additional super-partners. At high temperatures, the three forces of the standard model unify into one force. Theories that predict this unification are called "grand unified theories" or "GUTS" for short (Greene (1999), pp. 174-180).

The irresistible elegance and simplicity that seems to explain so much convinced many theorists that string theory is a unification of physics. In addition, there is strong evidence that string theory is finite and does not suffer from the "infinity problem."

Despite its positive features, string theory has detractors. There are five ten-dimensional superstring theories, none of which describes our Universe. All five theories sport six extra spatial dimensions not directly observed. The extra dimensions introduce complexities into the theory. The only feasible manner of describing these extra dimensions is to imagine them curled up into small, difficult-to-detect compactions. But, loosely speaking, the curling up can be accomplished in innumerable ways, leading to thousands of variations of string theory. None of the variants seem compatible with the standard model (Smolin (2007), § 10).

However, in 1995, Witten (2002) conjectured that the five superstring theories might be part of a more fundamental theory, which he called "M-theory." But, M-theory has yet to be completely defined. It attracts criticism for lacking predictive power, for being incomplete and untestable. Likely, it will require the development of a new mathematical language, which makes further progress on string theory difficult. The question, "What is string theory?" must have an answer at some point.

The M-theory conjecture had barely seen daylight before it was realized that newer higher-dimensional objects called "branes" must be introduced to create a feasible theory. Branes are entities to which the ends of strings can attach themselves. Including "branes" gave string theory added flexibility, including the ability to model black hole entropy.

Inspired by the holographic principle, in 1997, Maldacena conjectured that there might be a way of interpreting the extra spatial dimensions in terms of real-world physics. Referred to as the "AdS/CFT correspondence," it showed an equivalence between a string theory and gravity defined on one space and a QFT without gravity defined on the conformal boundary of that space. This discovery suggested that the quantum gravitational theory embodied in string theory is equivalent to a QFT with one less dimensional element, namely gravity. It was the first time string theory could be connected with real-world physics (Jones (2010), § 11).

But the AdS/CFT correspondence conjecture employs background geometries in anti-de Sitter spaces. Anti-de Sitter geometries have negative curvatures, which require a negative cosmological constant. But, so far, all observations indicate, fairly conclusively, that the cosmological constant associated with our Universe is positive. Whether or not the same conjecture will hold in model universes with a positive cosmological constant remains an open question (Smolin (2007), pp. 153-158).

All current supersymmetric string theories include "supersymmetry," which conjectures that all fermions have a bosonic partner. But none of these super-partners has, as yet, been experimentally observed. One possibility is that the super-partners are so massive it would take enormous energy to detect them. This situation suggests that either supersymmetry does not exist or the symmetry is broken somehow. If supersymmetry is ruled out as a fundamental part of Nature, it could end the future of string theory.

While introducing "branes" brought a welcome amount of flexibility to string theory, it significantly increased the number of background geometries upon which the string could live, further increasing the number of possible consistent string theories (Gross (2005)). However, at this point, it is unclear if any of the string theories can describe our Universe.

The critics of string theory argue that, with so many possible theories, there is no way of falsifying the theory. Therefore, they argue that string theory has a mythical character, will make no testable predictions, and, in the final analysis, is not science. To counter this view, some string theorists have endorsed the concept of "the landscape" of string theories (i.e., many mathematically consistent possible universes, some of which may exist). The landscape is their solution to the unfathomable number of possible string theories, where the anthropic principle is suggested as a way to identify which string theories might describe our Universe or our part of the Universe.

The anthropic principle is controversial. It is often viewed as unscientific because it is sometimes invoked as proof that a supernatural designer of the Universe exists or, in its weaker form, as an argument against the existence of a supernatural designer (Jones (2010), § 11).

John Schwartz and Michael Green provided strong evidence that string theory is finite, which means it should give finite answers to sensible questions and avoid the "infinity problem." But there is no general agreement on this point. Perturbation approximations, the primary tools employed by string theorists, generally have an infinite number of terms. In the final analysis, convincing proof must be given that string theory is finite (Smolin (2007), pp. 278-281). And so far, not all physicists agree whether this has been achieved or even if it is required.

Whether or not string theory becomes an acceptable quantum theory of gravity remains an open question. It shows great promise. But it must overcome many issues, not the least of which is answering the question, "What is string theory?" Recent progress has been slow, so watching and waiting has become a primary activity.

### 1.2.3 Third Future: A New, More Fundamental Theory

Many physicists subscribe to the idea that a new, more fundamental theory is the only road to unification. Frustrated with the quantum theory because "it doesn't make sense," many physicists search for an explanation that replaces the quantum theory.

But those who think along these lines generally have an "I'll know it when I see it" intuition rather than a well-thought-out approach to a new theory. Consequently, no serious "new theory" is capturing the imagination and support of most physicists. Quite the contrary, those who support this idea, for the most part, appear genuinely conflicted as to what approach is required. Lacking an acceptable alternative, physicists mainly cast an eye toward string theory.

One of the crucial questions is, "Do the ideas associated with Natural laws belong in a theory, or are our ideas merely representative of those laws? The question boils down to determining what our ideas are made of, if they exist. If ideas are the stuff of a material substance, then there should be little debate over whether they belong within the domain of physics. Presumably, material ideas would be governed by the same Natural laws that underpin the physics of other material things, brain waves, for example, that could, in some manner, be detected either directly or indirectly through sense experience.

Traditionally, philosophers have regarded ideas as non-material - not existing in space and time. And physicists are reluctant to include anything within the domain of physics that cannot be verified through sense experience, at least indirectly. Conventionally, things perceived through sense experiences are considered "material" - made of matter/energy. Beyond that, we are endowed with a mind, usually regarded as independent of and non-material to our "material" body. Physicists would generally expect experimental verification of theoretical suppositions, and it is unclear how experimental verification of a non-material thing could exist. Nevertheless, most physicists admit the existence of dark energy/matter that does not register on their detection devices. Such dark observations are verified by observing the behavior of observable matter near the dark substance, whatever it is.

The second question is, "Does an external world exist independently of our perception of it?" I think most physicists would answer in the affirmative. It would seem odd if physicists adopted the contention of Lord Berkeley, who denied the existence of an external world. What are physicists writing and prognosticating about if it isn't an external world?

The prevailing attitude amongst physicists is decidedly agnostic toward this question. For example, current interpretations of the quantum theory avoid the issue of whether or not an external reality exists - simply claiming that what's pertinent is whether a theory predicts the correct outcome of a measurement. This prognostication is precisely what the quantum theory suggests - the only reality is the one revealed after a measurement. Outside of that, knowledge of the unobserved world is out of reach; probabilities, but no reality, can be assigned to it.

If an external world exists, are our mental perceptions faithful representations of that world? If our perceptions are not faithful representations of an external world, what are they representations of?

The quantum theory gives probabilities for the possible realities that appear when an observation is made. These are the questions that this book will address. But, before focusing on these critical questions, it will be necessary to understand the issues faced in the theoretical attempts at unifying all of physics.

# Chapter 2: The Struggle for Unification

***"The microwave oven is the consolation prize in our struggle to understand physics."***

***– Jason Love***

## 2.0 Introduction

Chapter One discussed how physics is dominated by two remarkably successful but incompatible theories - the theory of relativity and quantum mechanics. From a historical perspective, examining how past events and discoveries drove modern physics toward a unified view will be instructive.

## 2.1 Einstein's Attempt at Unification

One of the initial attempts at a unified theory of physics came from Einstein. The pursuit lasted the last thirty years of his life. Over a period of about ten years, from 1905 to 1915, he formulated the special and general theories of relativity. The special theory corrected anomalies within Newtonian mechanics. His general theory produced a theory of gravity. Maxwell had shown that electricity, magnetism, and light were all manifestations of the electromagnetic field. However, the general theory of relativity left Maxwell's theory unaddressed (besides gravity, the electromagnetic force was the only other force known at the time), and Einstein regarded the omission as a "deficiency."

The physics of the twentieth century saw a movement towards a more personalized (local) interpretation of physical laws, where the point of view of the individual observer became increasingly dominant. The systems of the prior generations, Newtonian physics, and earlier attempts at describing the electromagnetic force assumed that the motions of objects could be described against an idealized stationary space, which served as a reference frame for all observers. A stationary space simplifies the job of describing how objects move. At this point, physics was governed by the law of "Galilean relativity." Galileo postulated that an observer, whether at rest or moving with a constant speed, would obtain the same results for all mechanical experiments. But when applied to electromagnetic experiments, Galilean relativity led to results that did not agree with observations. Consequently, arguments designed to justify the observational anomalies became increasingly indefensible, resulting in a revolution in physics.

The revolution saw two significant changes in perspective. First, physics became both localized and individualized. Before the arrival of the revolutionary ideas of Einstein and Maxwell, the gravitational and electric forces between two particles were described by an "action-at-a-distance" formulation. The force between two masses or two charges took effect instantaneously regardless of how far apart the masses or charges were separated in space. But such "action-at-a-distance" violated both common sense and modern theories. According to special relativity, nothing travels faster than light. Having a limit on how fast signals could propagate through space made it difficult to defend the idea of an instantaneous "action-at-a-distance."

The "action-at-a-distance" formulation, which maintains that physical influences are transmitted through empty space without any material or physical agency, was eventually replaced with a localized field concept in both the theories of electromagnetism and gravity. A field is a region throughout space in which two bodies separated in space exert a force on each other. Rather than a body, *A*, directly exerting a force on the body, *B*, body, *A*, creates a field in every direction around it, and the body, *B*, feels the effects of the field at its position in space. When a change occurs at the source, its effect propagates outward through the field at a constant speed. Its effects are felt after a finite amount of time. Each type of force (electric, magnetic, nuclear, or gravitational) has its own field; a body experiences the force only if it is also a source of that kind of field. If two bodies exert a mutual force, they possess potential energy that depends on their relative positions; energy is regarded as residing in the field and not in the bodies.

The individualization of physics significantly modified the job of the physicist. Before the theory of relativity arrived, Natural laws remained unaffected if observers moved with a constant speed relative to one another (Galilean relativity). Physicists did not worry that the clock of a stationary observer might run at a different rate than an observer in motion or that an observer at rest might obtain a different value of a distance measurement than an observer in motion. But experiments with light showed that Galilean relativity failed.

The special theory of relativity, developed in 1905, restricted theoretical prognostications to inertial reference frames, where observers are moving with a constant velocity relative to one another. It succeeded in removing the contradictions between Newtonian mechanics and Maxwell's theory of electromagnetism. Maxwell's theory maintained that light traveled at a finite speed. The theory of relativity postulated that all inertial observers would measure the velocity of light (in a vacuum) to be what Maxwell's theory said it was. This restriction meant that light signals would not necessarily reach all inertial observers simultaneously. The previously accepted idea of event simultaneity for all inertial observers was replaced by the "principle of relativity," which made the weaker claim that physical laws must be identical in all inertial reference frames.

This change in perspective produced consequences that seemed odd but true. Time became local. Observers in two different inertial frames would not necessarily agree on when an event occurred. Neither would they necessarily agree on where in space the event happened. What all inertial observers would agree on was the "space-time interval," expressed by the equation

$$x^2 + y^2 + z^2 + (ict)^2 = x'^2 + y'^2 + z'^2 + (ict')^2,$$

where $c$ is the speed of light in a vacuum and the spatial locations and times of an event are given by $(x, y, z, t)$ for one observer and $(x', y', z', t')$ for the other observer. While there might not be agreement on the exact location or the time of an event (i.e., $x \neq x'$ and $y \neq y'$, etc.), the two observers would agree on the space-time interval. A physicist would say that the space-time interval remains "invariant" in all inertial frames. Often referred to as "Lorentz invariance," the equation is considered a law of Nature subject to the constraint that it only holds in inertial systems. The redefinition of simultaneity resolved the conflict between Newtonian mechanics and the fact that light speed was a constant. It provided a conceptual justification for Lorentz invariance. Einstein successfully integrated two major fields of physics into what amounts to a conceptual unification (Sauer (2007)).

Special relativity demands a finite limit on the speed of light, which applies to any signal transmission, and this constraint conflicts with Newtonian gravitational theory. A gravitational interaction between massive bodies by instantaneous "action-at-a-distance" contradicted the postulate that signal transmissions could not exceed the speed of light. The conflict between Newtonian gravitation, conceptualized as "action-at-a-distance," and Maxwell's theory, conceptualized as a dynamic field propagating through space, led Einstein to the "principle of equivalence" (Sauer (2007)):

> *"We arrive at a very satisfactory interpretation of this law of experience, if we assume that the systems K and K′ are physically exactly equivalent, that is, if we assume that we may just as well regard the system K as being in a space free from gravitational fields, if we then regard K as uniformly accelerated. This assumption of exact physical equivalence makes it impossible for us to speak of the absolute acceleration of the system of reference, just as the usual theory of relativity forbids us to talk of the absolute velocity of a system; and it makes the equal falling of all bodies in a gravitational field seem a matter of course.*
>
> – Einstein, *1911*

Einstein elevated the equivalence of inertial and gravitational mass to the status of a general principle:

> *"As long as we restrict ourselves to purely mechanical processes in the realm where Newton's mechanics holds sway, we are certain of the equivalence of the systems K and K′. But this view of ours will*

> *not have any deeper significance unless the systems K and K′ are equivalent with respect to all physical processes, that is, unless the laws of nature with respect to K are in entire agreement with those with respect to K′. By assuming this to be so, we arrive at a principle which, if it is really true, has great heuristic importance. For by theoretical consideration of processes which take place relatively to a system of reference with uniform acceleration, we obtain information as to the career of processes in a homogeneous gravitational field."*

This observation culminated in the theory of general relativity. In Einstein's version, the principle asserts that in free fall, the effect of gravity is abolished in all possible experiments. General relativity then reduces to special relativity. His insight allowed the principle of relativity to be applied to non-inertial systems and provided a local field theory of gravity. It was local because gravitational effects would depend on where an observer was stationed within the gravitational field. And it was individualized because the time registered on the clocks of two observers would not agree if the gravitational potentials were not equivalent at the places where the observers were located.

Still, the success of this unification came at a price. The gravitational interaction was conceptualized as purely geometrical, involving the geometric warping of space and time. Einstein now believed that his geometrized theory of the gravitational field demanded unification with the electromagnetic field.

He faced a huge challenge in attempting to unify gravity with electromagnetism. The equations of general relativity are non-linear, whereas those describing electromagnetism are linear. Gravity, as it turned out, was not a force but a consequence of the warping of space and time by the presence of matter/energy. The differential equations describing gravitational effects require that space-time be modeled mathematically by continuous functions. Benjamin Franklin regarded electricity as a continuous fluid, but today, "electric charge" implies that electrons come in discrete packets generally referred to as "quanta." This meant that the smooth continuous non-linear geometrically based field equations of general relativity must somehow be transformed into chunky quantized entities. And it was unclear how the "quantization" of space-time could be accomplished. The problem boiled down to describing how discrete particles arose from a smooth space-time.

In 1921, Theodor Kaluza purposed adding a fifth dimension to space-time. He was able to show that his theory contained four-dimensional general relativity in the presence of an electromagnetic field. But the extra dimension was bothersome. Where did it come from, and why was it not observable? Finally, in 1926, Oscar Klein proposed that the fifth dimension was curled up into a circle of a tiny radius that would not be observable except at very high energies (Kaluza (1921), pp. 966–972, Klein (1926), pp. 895–906).

In a sense, Einstein became smitten by the Kaluza-Klein approach. As a minimal extension of his general theory, he envisioned gravity and electromagnetism (space-time and matter) unified into a purely geometrical theory. He had to show that particles emerged as non-singular solutions of matter fields (Gross (2005)). It is interesting to note many of his colleagues, at the time, considered Einstein's pursuit of extra dimensions a waste of time. It is ironic and somewhat of a testament to Einstein's foresight that, of the current theories considered serious contenders for a unification of physics, most sport numerous compactified spatial dimensions. Nevertheless, the quantization of matter fields from the gravitational field remains an unsolved problem.

### 2.1.1 The Emergence of the Quantum Theories

Meanwhile, physics was experiencing a revolution of a different sort. In 1894, Max Planck began studying the effects of black-body radiation. A black body is an idealized object capable of perfect emission and absorption of radiation. Planck was interested in discovering how the intensity of the electromagnetic radiation emitted by a black body depends on the radiation's frequency and the body's temperature. Planck knew that the current theory, embodied in the classical "Rayleigh-Jeans law," agreed with experimental results at low frequencies but failed to do so at high frequencies.

Furthermore, the law predicted that infinite temperature increases resulted in infinite energy emissions (Kragh (2000)). But experiments showed that this was not the case. Instead, a body would heat up to a specific temperature, then absorb as much radiation as it emitted.

Planck came to a better solution by realizing that the interplay of emission and absorption in the black body kept the energy from shooting to infinitely high levels. His equations employed an exponential damping function, and he made statistical arguments, which predicted the probability that the black body would emit or absorb an amount of radiation. Consequently, his approach was in better agreement with experimental results. But to achieve an acceptable solution, Planck assumed that electromagnetic energy was emitted only in discrete energy packets, $E$, which came in multiples of an elementary unit, $E = h\nu$, where $h$ is a constant (subsequently known as "Planck's constant") and $\nu$ is the frequency of the radiation (Beiser (1967), § 10.5).

Planck's discovery would lead to a revolution in physics. His "discrete packets of energy" became a fundamental feature of the new physics of the "quantum." After Planck's discovery, more and more phenomena, not only radiant energy but material energy as well, required explanations that involved the "discrete energies" initially described by Planck.

### 2.1.2 The Photoelectric Effect

Experiments had shown that electrons were emitted when light fell on a thin piece of foil. Einstein explained this phenomenon by assuming that light acted like a stream of particles, which he named "photons." Called the "photoelectric effect," the hypothesis gained a reasonable explanation only if it was assumed that the photons came in Planck's discrete chunks of energy ($E = h\nu$) (Beiser (1967), § 2.1). But the chunkiness of light contradicted Maxwell's theory, which maintained that light propagated as an electromagnetic wave. In 1923, Arthur Compton published a paper confirming the photoelectric effect for which he won the 1927 Nobel Prize (Taylor (2015), pp. 136-9).

Unwittingly, Einstein had contributed to the irresistible force of the new quantum revolution - a revolution he would later openly detest.

### 2.1.3 Matter Waves

Two decades later, in 1924, Louis de Broglie proposed that not only did radiation exhibit particle properties, but particles exhibited wave properties. In his doctoral thesis, "*Recherches sur la théorie des quanta*," based on the work of Planck and Einstein, he introduced the theory of electron waves. His research led to the "wave-particle duality" theory of matter, where the momentum of a particle depended on its wavelength. The doctoral thesis examiners, unsure of the material, passed his thesis to Einstein for evaluation. Einstein endorsed the wave-particle duality proposal wholeheartedly. De Broglie had established a new concept in physics, "matter waves," uniting the physics of energy (wave) and matter (particle). He won the Nobel Prize in physics in 1929 for his work (Beiser (1967), § 3).

With every particle with mass, $m$, and velocity, $v$, a real wave must be associated, related to the momentum by the equation

$$\lambda = \frac{h}{p} = \frac{h}{mv}\sqrt{1 - \frac{v^2}{c^2}},$$

where $\lambda$ is the wavelength, $h$ is Planck's constant, $p$ is the particle's momentum, and $c$ is the speed of light in a vacuum. De Broglie's equation was validated by the electron diffraction experiments of Davisson and Germer (Beiser (1967), § 3).

The equation above is interesting because if $v = c$, $\lambda\ (the\ wavelength) = 0$. But light, which travels at the velocity, $c$, comes in all different wavelengths, not the zero wavelength the equation predicts. This observation suggested that de Broglie's equation only applied to particles that have a mass. And it further supports the notion that particles with a mass cannot obtain the speed of light. Moreover, if $v = 0$, the wavelength of the particle is infinite. A massive particle at rest does not jiggle.

On the other hand, in empty space, the photon (light particle) moves at speed, $c$, and its energy and momentum are related by the relativistic relation ($m = 0$)

$$E^2 = p^2c^2 + m^2c^4.$$

The energy and momentum of a photon (a massless particle) depend only on the frequency ($\nu$) or, inversely, its wavelength ($\lambda$): $E = pc$, where $p = \hbar k$, $k = 2\pi/\lambda$ is the "wave vector," and $\hbar = h/2\pi$ is the reduced Planck's constant (more on this later). The magnitude of the momentum is

$$p = \hbar k = \hbar \frac{2\pi}{\lambda} = \frac{h}{\lambda}.$$

The relationship between a massless particle and a particle with mass plays an essential role within the standard model of particle physics. The Yang–Mills theory, the theoretical part of the standard model, is a generalization of Maxwell's theory of electromagnetism, where the chromo-electromagnetic field carries charges. As a classical field theory, it has solutions that travel at the speed of light, so its quantum version should describe massless particles (gluons). However, the theory of color confinement in quantum chromodynamics permits only bound states of gluons, forming particles with mass. This problem is called the "mass gap" (Jaffe). The mass gap is still an unsolved problem in quantum physics.

De Broglie's fundamental equation relating the momentum of a particle to its wavelength was a precursor to the uncertainty principle. De Broglie introduced the idea of a "wave packet," which is a "burst" or "envelope" of wave action that travels through space as a unit (see figure below).

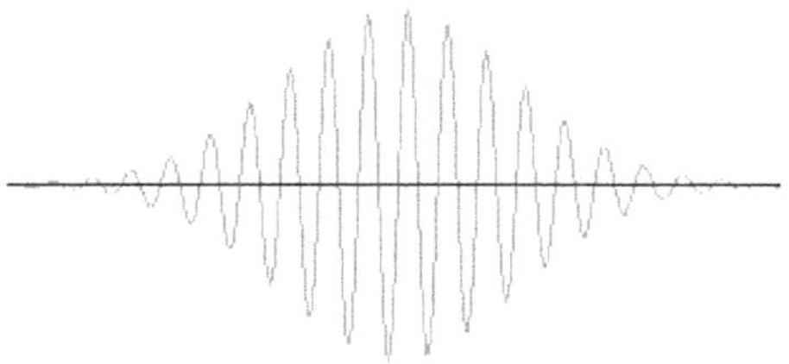

The packet is a set of component waves of various energies that interfere constructively over a small spatial region but destructively outside that region. The size and shape of the wave packet may remain constant or change while propagating. De Broglie showed that wave packets display many of the properties of the electron - a particle (Beiser (1967), § 3).

De Broglie's equation, $p = h/\lambda$, implied that confining the wave packet to a tiny area ($\lambda \ll 1$, $p \gg 1$) made it difficult to measure its exact momentum and position simultaneously as the wave packet flashed by a suitable measuring device. The error in measuring the exact location of the wave packet was related to the wavelength by $\lambda \propto \Delta x$, where $\Delta x$ is the error in the position measurement. Analogously, $p \propto \Delta p$, where $\Delta p$ represents the error in the momentum measurement. Plugging these two values into de Broglie's equation:

$$p = \frac{h}{\lambda} \rightarrow \lambda p = h \rightarrow \Delta x \Delta p \geq h,$$

which says that the error in simultaneously measuring momentum and position is at least as great as $h$ (Planck's constant). In the coordinate representation of a wave packet (such as the Cartesian coordinate system), the position of the wave packet is a function of $\lambda$. Therefore, a spatially narrower wave packet allows a better position measurement at the expense of the spread in the wave's momentum. This trade-off between the ambiguity in position and the spread in momentum is an example of the "Heisenberg uncertainty principle" (Beiser (1967), § 3).

### 2.1.4 Signs of a Brewing Controversy

From a philosophical standpoint, de Broglie thought a real wave (having a direct physical meaning) was associated with the particles. The wave aspect of matter was formalized by the "wave function" in the Schrödinger equation. Subsequently, Max Born gave the Schrödinger equation a purely statistical interpretation. His explanation became part of the Copenhagen interpretation of quantum mechanics, which postulates that the wave function carries no physical meaning but only gives an appearance of wave behavior to matter without making actual physical waves appear. But until the end of his life, de Broglie believed in a real, physical interpretation of matter waves. Wrote de Broglie:

> *"...the particle must be the seat of an internal periodic movement and that it must move in a wave in order to remain in phase with it was ignored by the actual quantitative physicists [who are] wrong to consider a wave propagation without localization of the particle, which was quite contrary to my original ideas."*

David Bohm (1993) later refined de Broglie's ideas. The de Broglie-Bohm theory is a non-local hidden variable deterministic theory whose predictions agree with the quantum theory. The de Broglie-Bohm theory is today the only interpretation giving actual physical status to matter waves. But, since it does not go further in its predictions than the Copenhagen interpretation, it is largely ignored.

The sentiments expressed by de Broglie would later erupt into a full-fledged ideological battle that still exists somewhat today. On the one side are the relativists who maintain that the quantum theory, as understood through the Copenhagen interpretation, does not make sense and should be replaced or reinterpreted. On the other side are those who, more or less, support the Copenhagen interpretation or some variation of it.

### 2.1.5 The Development of the Atomic Model

By this time, physicists began contemplating how these ultra-small particle waves of matter should be described ontologically. Mostly, they concluded that the fundamental constitution of matter consisted of small indivisible entities (atoms). The concept of an atom dates from the time of Aristotle, who thought the Earth was made from four essential elements (earth, wind, fire, and water). Contemporaries of Aristotle (Democritus, Aristarchus, and Archimedes), to one extent or another, rejected his principles in favor of other ideas (atomism and the idea that the planets revolved around the Sun) that were closer to modern conceptions. But "The Church" so favored Aristotle's ideas that they made him the "Official Philosopher of the Roman Catholic Church." At the time, the Church played a dominant role in all aspects of human life. It dictated which ideas were worthy of being "true" ideas. It was heresy even to conduct an empirical test or experiment that challenged the Church's official position on a topic, Galileo being the most famous victim of the Church's inquisitions. But eventually, the Church's influence waned, and the Enlightenment period led to thinking of a more "scientific" nature, which became more persuasive.

In 1827, while examining grains of plant pollen suspended in water under a microscope, Robert Brown observed that the minute particles executed a continuous jittery motion. He then observed the same motion in particles of inorganic matter, enabling him to rule out the hypothesis that the effect was life-related. Although Brown did not provide a theory to explain the motion, it eventually became known as "Brownian motion" (Pearle (2010)).

By the time Einstein took up the subject of Brownian motion in 1905, the positivist philosophical crusade had emerged. The positivists argued that the multitude of metaphysical arguments sprinkled throughout physics was the primary source of theoretical errors. The only concepts that should find a place in physics were those verified through sense experience. For example, the positivists, at the time, supported the view that atoms did not exist because they were, for all intent and purposes, unobservable. But Einstein was able to determine the size of atoms and how many atoms there were in a mole - the molecular weight (in grams) of a gas per Avogadro's law. Einstein's paper is widely credited with proving that atoms were measurable objects that fit the requirements of positivist thinking at the time (Einstein (1956)).

Sir J. J. Thomson, the 1906 Nobel Prize winner for the discovery of the electron, was one of the first to propose a model of the atom. His model was composed of electrons (Thomson called them "corpuscles") surrounded by a soup of positive charges that balanced out the electron's negative charges, somewhat like negatively-charged "plums" surrounded by a positively-charged "pudding." The many arrangements of the electrons in the pudding constituted the numerous possible atoms (Beiser (1967), pp. 86-88).

In 1909, under the direction of Ernest Rutherford, Geiger and Marsden tested the Thomson model by bombarding gold foil with a beam of alpha ($+2e$) particles. Since the electrons were assumed uniformly distributed throughout the "pudding," the alpha particles should have been deflected by, at most, a few degrees. But the experiment showed that a small percentage of the alpha particles were deflected through angles much larger than $90^{\circ}$, suggesting that powerful forces were present to enable such marked deflections (Beiser (1967), pp. 86-88).

The results led Rutherford to reject Thomson's model in favor of a miniature solar system model, where a large positive electric charge was concentrated into a small (as compared with the size of the atom) central region. Still, outside this "central region," the atom was mostly empty space. Instead, the large positive core was orbited by smaller (in size), oppositely charged electrons, rendering the atom roughly electrically neutral (Beiser (1967), pp. 87-96).

The deflection experiments proved useful in determining the nuclear structure of many materials. The deflection patterns varied depending on the type of material, providing insight into the configuration of the atoms that made up the material. The experiments also showed that electric charge came in discrete units, providing further evidence that electric charge was a quantum phenomenon (Beiser (1967), § 4.3).

Using purely Newtonian methods, Rutherford calculated the energy required for the electron to remain in orbit around the nucleus of a hydrogen atom (an atom with one orbiting electron). He also calculated the radius of the electron orbit. His calculations agreed with estimates made in other ways (Beiser (1967), § 4.5).

But the Rutherford model didn't square with the classical theory of electromagnetism, which predicted that accelerating electrons emit radiation, which would cause the electron to lose energy. Electrons in circular orbits were, in effect, accelerating and should lose energy, eventually spiraling into the nucleus. But atoms were stable. The "spiraling into the nucleus" effect rarely happened. Reportedly, Rutherford was on the verge of scrapping the miniature solar system atomic model until Niels Bohr came to the rescue.

Subsequently, Bohr joined Rutherford at Manchester University as an understudy. Bohr noticed that when passing an electric current through a gas, the atoms emitted radiation in a spectrum of only specific wavelengths. Moreover, the wavelengths came in a discrete spectral series, uniquely identifying each element. The first such series was found by J. J. Balmer in 1885 when he studied the visible part of the hydrogen spectrum. Later, the invisible spectral series of hydrogen was also developed. The equations that represented each series all exhibited the mathematical form,

$$\frac{1}{\lambda} = R\left(\frac{1}{m^2} - \frac{1}{n^2}\right),$$

where $n$ and $m$ are integers, $\lambda$ the wavelength, and $R$ is "Rydberg's constant." Bohr hypothesized that the electron could circle the nucleus indefinitely without radiating energy provided that the orbit contained an integral number of de Broglie wavelengths. Using the de Broglie theory, the electron wavelength could be used to calculate the orbital speed required to balance the electrostatic force of the nucleus. The various orbits represented the permitted energy levels the electron could assume. Bohr was able to calculate the energy levels and relate them to the spectral series of the hydrogen atom in almost perfect agreement. His discovery was a startling development and a monumental achievement. Bohr had saved the Rutherford model. Atoms would remain stable, avoiding the destructive collisions between the spiraling electrons and the oppositely charged nucleus. Bohr also showed that when electrons either absorb or lose energy, they jump from one energy level to another in discrete units, $E_i - E_f = h\nu$, where $E_i$ is the initial energy state and $E_f$ is the final state. Electrons lost energy in the form of radiation (Beiser (1967), § 5) that came in discrete units of $h\nu$.

## 2.2 The New Physics

The Rayleigh-Jeans law is deducible from Maxwell's electromagnetic theory[1] but provides an inaccurate description of the emission and absorption of radiation. This fact, coupled with the discovery that electrons were confined to discrete orbits as described by Bohr's model of the atom, the theory of de Broglie's matter waves, and Einstein's analysis of the photoelectric effect, provided overwhelming evidence that, in some sense, classical theories harbored fundamental flaws.

The Bohr model correctly predicted the spectral series of the hydrogen atom. But its predictions could not be extended to more complex atoms having more electrons. Moreover, it could not explain why specific transitions of energy levels were more likely to occur than others, account for slight differences in certain spectral lines, nor adequately account for the formation of molecules (Beiser (1967), p. 125). Clearly, a broader theory was required.

### 2.2.1 The Wave Function

The broader approach was developed by Erwin Schrödinger and Warner Heisenberg (Schrödinger (1926), pp. 1049-1070, Heisenberg (1927), pp. 172-198). Schrödinger formulated his version of quantum mechanics in 1925, for which he received the Nobel Prize in 1933. His equation would become to quantum mechanics what $F = ma$ is to Newtonian mechanics. In parallel, Heisenberg, Max Born, and Pascual

---

[1] For a complete derivation see: Bohm, D., *Quantum Theory*, D. Bohm, 1951, pp 7-17.

Jordan developed the matrix formulation of quantum mechanics, for which Heisenberg received the 1932 Nobel Prize. Schrödinger would later show that his and the Heisenberg formulations were equivalent.

Quantum systems are described by a wave function, traditionally signified by the Greek letter, $\psi$. The wave function represents a solution to the Schrödinger equation. In other words, $\psi$ represents the possible "states" that a quantum system can adopt at any moment. It contains all the information known about a system. Of questions asked about a system, answers can only be obtained if the information is contained within $\psi$.

Like classical mechanics, quantum mechanics involves relationships between observable magnitudes such as "velocity" and "position in space." But classical physics takes for granted that specific measurable quantities, such as a particle's mass, velocity, acceleration, etc., can be known simultaneously. But the uncertainty principle limits how much can be known about specific quantities. Moreover, unlike classical physics, where a series of measurements of a single observable gives the same value after each measurement, quantum measurements are generally unpredictable. Consequently, it is only possible to give an average value of a series of measurements (Greensite (2003), § 2).

Under the current and most accepted understanding of quantum mechanics, $\psi$ has no physical interpretation. On philosophical grounds, this seems paradoxical. How can a fundamental concept within physics have no physical reality? But giving a physical interpretation to $\psi$ is fraught with difficulties. Specific paths for particles, like electrons, cannot be found. Moreover, it is not possible to obtain repeatable measurements in every case. For these and other reasons, the inventors of quantum mechanics were driven toward the paradox of the quantum wave function as a primary concept.

The uncontrollable nature of quantum processes forced its developers to work in terms of probabilities rather than specific physical outcomes. In classical physics, if the initial values are known at some point in space and time, future values can be predicted with certainty. But $\psi$ can only be associated with probabilities.

While given no physical interpretation, a valid $\psi$ must possess specific characteristics: $\psi^2$ must be positive and proportional to the probability of finding, for example, a body at a given location at a given time. Requiring that $\psi^2$ be positive removes any chance that negative probabilities occur since such probabilities are meaningless. If all the values of $\psi^2$ are added up, the sum must be finite. An electron, for example, must inhabit a location at all times within the quantum system under consideration. All values of $\psi^2$ must be real (no complex numbers). If $\psi^2$ were complex, a system could be found in more than one state at a time, which never happens. A valid $\psi$ must be a single-valued function of location and time. It would not do, for instance, if a body could have a 30% and a 50% chance of simultaneously occupying a given location. Finally, the partial derivatives of all quantum mechanical wave functions must exist, a mathematical statement determining the requisite smoothness required for $\psi$ to be a legitimate wave function. Roughly speaking, its values cannot jump randomly but must comply with what mathematicians call "continuity." Continuity is a classical concept, which might seem odd since quantum processes are unpredictable. Ironically, Quantum processes are smooth and gradual in the world of $\psi$. However, as soon as an interaction with a measuring device occurs, $\psi$ ceases to be a factor, and, in its place, a real outcome appears. Jumping from the mathematically smooth world of the wave function to "after a measurement is taken" is called "collapsing the wave function." It is in this sense that quantum processes jump. Before a measurement, a quantum system can be in many possible states simultaneously, but when a measurement is taken, the system will be found in just one of those states. Keep in mind, however, that although wave functions are continuous (smooth) in the mathematical sense, they are probability distributions.

### 2.2.2 Operators

Nothing in the quantum theory induces a real outcome unless an observation is made. Until then, outcomes are described in terms of the possibilities associated with the wave function. The jump from the unseen world of the wave function to a real outcome is represented by an "operator."

An operator modifies the wave function. In quantum mechanics, operators act on $\psi$, expressed by the following equation, $O\psi = \lambda\psi$. Here $O$ represents the operator, $\psi$ is the wave function, and $\lambda$ is called an "eigenvalue." In quantum mechanics, an operator must be a special type called a "Hermitian operator." Eigenvalues are real numbers. When the operator ($O$) acts on the wave function, it signifies that an observation or measurement has taken place. At that point, the right-hand side of the equation becomes a reproduction of the wave function multiplied by one of its eigenvalues. In essence, the operator has extracted information from the wave function. The eigenvalue represents the outcome of a measurement.

There is an associated operator with every physical observable (position, momentum, energy, angular momentum ...). More than one operator can operate on a wave function, signified by $AB\psi$, where $A$ and $B$ are the operators, which indicates that $B$ operates on $\psi$ first, followed by $A$. $BA\psi$ signifies the opposite.

The order in which the operators act on the wave function is significant. This is expressed mathematically by the commutator, $[A, B] = AB - BA$, where $A$ and $B$ are operators. If $[A, B] = 0$, the order in which the operators are applied to the wave function is immaterial. But if $[A, B] \neq 0$, then the order does matter, and applying $A$ then $B$ gives a different outcome than applying $B$ then $A$. What the commutator signifies is whether the two observables, represented by the operators, $A$ and $B$, can be measured simultaneously. If $[A, B] = 0$, the answer is yes; otherwise, the answer is no.

Operators express the unpredictable nature of quantum processes. Suppose a particle is in an unknown state, its momentum measured and found to be $p$. If the undisturbed particle's momentum is measured again, quantum mechanics guarantees that the answer will still be $p$. A particle in an eigenstate of momentum has a definite momentum, $p$, with probability, 1. However, suppose the second measurement measures the particle's position. In this case, the answer will be completely uncertain. The particle is equally likely to be in any possible position. A particle in a momentum eigenstate doesn't have a particular position. This is an example of the uncertainty principle, which requires that if the momentum of a particle is known precisely, then its position becomes completely indeterminate and vice-versa. Position and momentum operators in quantum mechanics do not commute. Therefore, it is impossible to obtain exact position and momentum measurements simultaneously.

Again, if an unperturbed particle's momentum is measured with the result, $p$, the same measurement can be made 100 times successively with the same result each time. If position is measured on the 101st measurement, the result will be unpredictable, but a result obtained nonetheless. The particle will be found somewhere. Suppose on the 102nd measurement, momentum is measured again. In this case, the momentum measurement will be unpredictable. When the position was measured, it forced the wave function to jump to an eigenstate of position. It is no longer in an eigenstate of momentum. The position measurement erases any memory of the momentum measurement. At that point, a momentum measurement becomes probabilistic.

Instead of measuring position, suppose kinetic energy and momentum are measured. In that case, the momentum measurement would still be $p$. The reason is that momentum and kinetic energy have the same set of eigenstates, so a measurement of the kinetic energy doesn't erase the momentum information. The two operators, momentum, $M$, and the kinetic energy operator, $K$, are examples of commuting operators ($[M, K] = 0$), which can be in the same eigenstate simultaneously, and hence, can be measured simultaneously.

### 2.2.3 Quantum States

Classical equations make deterministic predictions of the future status of physical systems. If the requisite number of initial conditions is known, classical physics predicts the system's future state with certainty. Classical systems can be divided into wave-like or particle-like phenomena. If the phenomenon under consideration is particle-like, particle-like equations describe the system. The same applies to wave-like phenomena (Bohm (1951), § 2).

But, as de Broglie and others proposed, submicroscopic particles like an electron have a dual nature. Particles display both particle and wave-like characteristics depending on their environment. From a classical standpoint, this creates a conceptual paradox. At the quantum level, there is no way of determining whether a particular phenomenon is wave-like or particle-like before collapsing the wave function. The situation implies that there is no way of selecting which types of equations (wave-like or particle-like) are applicable in describing the physical system of interest (Bohm (1951), § 2).

The solutions to Schrödinger's equation for electron energy levels in the hydrogen atom agreed with what the Bohr model predicted. Moreover, the equation could be extended to more complicated atoms and even to particles not bound in atoms. Despite its success, interpreting what "matter waves" represent remains controversial. How was it possible that Schrödinger's equation could describe both a wave, which is decentralized, spread out over an entire space, and a particle, where its center is located at a point? Schrödinger believed the intensity of the wave at a point in space represented the "amount" of the electron that was present at that point. In other words, the electron density was spread out, rather than localized. But this contradicted what is experienced, particles localized.

Max Born's statistical interpretation of Schrödinger's equation prevailed in the final analysis. For example, he argued that the wave associated with the electron was not tangible but a wave of probability that determines the likelihood of finding it at a given location at a given time. This made it possible to conceive of a particle as localized in space and time. The probability of finding the particle at a particular location would be large in the regions where the "matter wave" had a large amplitude. If the amplitude was small in that region, the probability of finding the particle there would be negligible.

Schrödinger's equation, like Newton's, $F = ma$, allows successful predictions, but in this case, the predictions are probabilistic. This interpretation was the only reasonable way, it seemed, to describe an electron, for example, that seemed to display both wave-like and particle-like characteristics. The electron is described as spread out over an entire space until it interacts with a measuring device, at which point, it becomes localized.

But accepting the "Born interpretation" meant rejecting the idea that an electron physically exists at a point in space and moves along a definite path. Instead, the electron could only be described, before taking a measurement, as a cloud of probabilities. Schrödinger's equation describes the evolution of these probabilities in time.

Schrödinger's equation has a time-dependent version describing progressive waves (i.e., progressive probabilities). It also has a time-independent form, which describes the probabilities associated with standing waves. Often called stationary waves, standing waves only allow specific wave amplitudes. For example, suppose the time-independent Schrödinger equation describes a simple pendulum. A pendulum can be thought of as a conserved energy system, assuming no energy is lost from the system due to friction or other external forces. On the downswing, the pendulum loses potential energy but gains kinetic energy. On the upswing, it loses kinetic energy but gains potential energy. Both classical and quantum pendulums conserve energy. But, described in classical terms, the kinetic energy of the pendulum can assume all possible values, including zero. But a quantum mechanical pendulum (usually called a "harmonic oscillator") can only assume specific values, not including zero. The time-independent Schrödinger equation predicts that only certain discrete energy levels materialize. An important distinction between quantum and classical physics is the discrete nature of quantum phenomena. The fact that quantum harmonic oscillators have no zero energy states is a direct consequence of the uncertainty principle.

Moreover, a classical pendulum is perfectly repeatable. On the upswing, it loses kinetic energy but gains potential energy up to a maximum potential of, say, $A$. It never exceeds $A$. But quantum pendulums are not perfectly repeatable. Each arm swing may reach a level of either more or less than $A$. It is only possible to think of $A$ as an average.

The time-independent Schrödinger equation describes the hydrogen atom more completely than the Bohr model. A hydrogen atom is a three-dimensional object, so its description requires a three-dimensional time-independent Schrödinger equation. Fortunately, without much loss of generality, the three-dimensional wave function can be mathematically separated into three one-dimensional wave functions treated individually (Beiser (1967), § 7.2).

The first wave function, $\psi_n$, predicts the probability that the electron occupies one of the discrete energy levels the electron can assume. Schrödinger equation predictions align with the Bohr model, which significantly validates Schrödinger's approach (Beiser (1967), § 7.3). Energy levels are typically signified by the number, $n$ $(n = 1, 2, 3, \dots)$.

The second wave function, $\psi_l$, predicts the probability of obtaining specific values of the orbital angular momentum of the electron. Angular momentum, like energy, comes in discrete quanta, which can assume only the values, $l = 0, 1, 2, \dots, (n-1)$. Quantized angular momentum was assumed within the Bohr model, but Schrödinger's equation predicts it, further validating Schrödinger's approach.

From this, the idea of a "quantum state" was spawned. Scientists began using convenient codes to describe quantum states. For instance, the levels of angular momentum are coded with letters

$$l = 0,1,2,3,4,5,6 \rightarrow 0 = s, 1 = p, 2 = d, 3 = f, 4 = g, 5 = h, 6 = i.$$

For example, a $2s$ state signifies that the electron is at energy level "two" and has "zero" orbital angular momentum. Whereas an atom with $n = 4$ and $l = 2$, is in a $4d$ state (Beiser (1967), § 7.5).

The third wave function, $\psi_{m_l}$, predicts the probability that the electron's orbital angular momentum obtains a specific direction. The direction of orbital angular momentum is quantized with respect to an external magnetic field. The magnetic quantum number, $m_l$, specifies the direction of $l$. The possible values are $m_l = l, \dots, -l$. In other words, if an electron is put into an external magnetic field, the third wave function predicts the orientation of its orbital angular momentum (Beiser (1967), § 7.6).

An atom in a state of principle quantum number, $n$, breaks up into several sub-states when submerged in an external magnetic field. Consequently, the energies are slightly more or slightly less than the energies of the $n$ state in the absence of such a field. The splitting of the spectral lines in the presence of an external magnetic field is called the "Zeeman effect" (Beiser (1967), § 7.6).

The Bohr model of the hydrogen atom predicted that the electron could always be found in the equatorial plane at $r = n^2 r_0$, where $r_0$ represents the innermost orbit of the electron. But Schrödinger's approach modifies Bohr's model in two important aspects. First, no definite values for $n, l, m_l$ could be given only relative probabilities for finding the electron in various states. Secondly, it was impossible to think of the electron as moving around the nucleus. Finally, since the time-independent Schrödinger equation describes the hydrogen atom, there is no way of describing the electron's position as a function of time (Beiser (1967), § 7.7).

### 2.2.4 Spin and Relativistic Quantum Mechanics

Despite the elegance with which the three-dimensional Schrödinger's equation accounts for certain aspects of the hydrogen atom, it cannot approach an adequate description of atoms without the concept of "electron spin" and the associated "exclusion principle." In addition, experiments showed that the spectral lines associated with the hydrogen atom consisted of two very minute lines. Referred to as the "fine structure" of the hydrogen atom, this spectral line separation is not predicted by the Schrödinger approach (Beiser (1967), § 8).

Secondly, the magnetic quantum number, $m_l$, accurately explained the "normal" Zeeman effect, but many elements displayed an anomalous Zeeman effect, where more orientations for orbital angular momentum were found than Schrödinger's theory predicted (Beiser (1967), § 8.1).

Goudsmit and Unlenbeck (1926, pp. 264-265) proposed that the electron possessed an intrinsic angular momentum that accounted for the fine structure and the anomalous Zeeman effect independent of orbital angular momentum. Goudsmit and Unlenbeck had in mind the classical idea of a charged particle spinning around its axis. Since the electron was charged, the rotation produced a magnetic moment opposite in direction to its orbital angular momentum.

However, the classical idea in line with what Goudsmit and Unlenbeck had proposed proved conceptually infeasible. Pauli pointed out that if the electron possessed the intrinsic spin that Goudsmit and Unlenbeck advocated, the rotation speed would exceed the speed of light. This observation put the concept at odds with the theory of relativity, where nothing travels faster than light. But in 1928, Dirac showed, based on a relativistic quantum-theoretical treatment, that the electron must have the intrinsic angular momentum and magnetic moment attributed to it by Goudsmit and Unlenbeck. Hence, electron spin became a purely quantum mechanical phenomenon with no classical analog (Ohanian (1986)).

Electron spin comes in only two discrete values: $\pm(\hbar/2)$. The idea of "electron spin" successfully explained the fine structure of atoms, the anomalous Zeeman effect, and many other atomic effects as well.

While Schrödinger's equation could explain the electron configurations in the hydrogen atom, it failed to do so for the more complex atoms. Atoms with atomic structures differing by just one electron displayed vastly different chemical properties. While one of these atoms might be a gas, the other could be a metal. It was hard to explain why the chemical properties of atoms could change so abruptly with only a minute change in the number of electrons and protons (Beiser (1967), § 8.2).

In 1925, Wolfgang Pauli found that each electron must have a different set of atomic numbers ($\{n, l, m_l, m_s\}$), where $m_s$ is associated with electron spin. Pauli noticed that specific state transitions were never found in all atoms but hydrogen. Given that an electron had atomic numbers, $n, l, m_l$, the forbidden transitions all had the characteristic that their spins could not add to 1 or $-1$. This observation meant that two electrons with the same quantum numbers, $n, l, m_l$, could not have the same spin direction (Beiser (1967), § 8.2).

Moreover, electrons and all elementary particles possess spin (some are spin-0). This was a monumental discovery. Ultimately, it allowed physicists to distinguish between matter and the forces that act on matter. Matter consists of spin-1/2 particles that obey the Pauli exclusion principle, which says that two or more identical particles with half-integer spins cannot simultaneously occupy the same quantum state. Such particles are called "fermions."

On the other hand, the force-carrying particles have integer spin, usually $0, 1$, or 2, and are called "bosons." According to the standard model of particle physics, bosons and fermions are the only types of particles in Nature. Bosons are not subject to the Pauli exclusion principle and can exist in the same quantum state.

With the help of Schrödinger's equation and the Pauli exclusion principle, atoms could be codified and classified into groupings with similar chemical properties. The coding took the form of a "subshell" (i.e., $nl^e$, where $n$ is the principal quantum number, $l$, is the orbital quantum number, and $e$, is the number of electrons in the subshell) (Beiser (1967), § 8.3-4). For example, $1s^2 2s^2 3p^6 3s^1$ is the codification for a sodium atom with its four subshells.

In 1871, Mendeleev formulated an empirical law called the "periodic law of the elements." He arranged the elements based on their atomic weight. Combining the periodic law with the more recent codification scheme eventually led to the modern-day "periodic table of the elements." For years, the periodic table of about 92 elements represented the fundamental building blocks of Nature. The development of the standard model reduced those 92 elements to about 17.

### 2.2.5 The Particle Zoo

By the mid-1960s, only the electron could be explained by a relatively satisfactory theory. Then, in 1928, Dirac discovered a wave equation that described elementary spin-1/2 particles, such as electrons, consistent with the principles of quantum mechanics and the theory of special relativity. In addition, it accounted for the fine structure of the hydrogen spectrum and implied the existence of a new form of matter, "antimatter," - later discovered experimentally. Dirac's explanation of spin as a consequence of the union of quantum mechanics and special relativity, and the eventual discovery of the positron (antimatter), represents one of the great triumphs of theoretical physics (Hatfield (1989)).

However, aspects of Dirac's theory were problematic. Nothing could prevent an electron in a positive-energy state from decaying into a negative-energy state by emitting photons (energy). But experiments showed that electrons did not act this way (Martin (2009)). To cope with the problem, Dirac envisioned the vacuum state as consisting of an infinitely large quantum state, where all negative-energy electron states are occupied. Since the Pauli exclusion principle forbids electrons from occupying the same state, any additional electron must occupy a positive-energy state. This "vacuum state" was eventually deemed the "Dirac sea." When the negative-energy states were incompletely filled, each unoccupied state created "a hole" that acted like a positively charged particle. Dirac initially thought the hole might be the proton, but Hermann Weyl pointed out that the hole should have the same mass as an electron, whereas the proton was over 1800 times heavier. The hole was eventually identified as the positron (antielectron), experimentally discovered in 1932 by Carl Anderson. He received the 1936 Nobel Prize in physics for the discovery (Beiser (1967), § 15.1).

However, conceptually, Dirac's idea of a vacuum state implied infinitely many unobservable negative electrons filling all of space, which exactly canceled the infinity of positive energy electrons (Beiser (1967), § 15.1). Most physicists at the time showed little inclination toward theories that made unobservable metaphysical declarations.

Eventually, Dirac's theory was replaced by a "QFT," which advocated the more convincing approach of treating the positron as a particle rather than the absence of a particle. QFT redefined the vacuum state as a state where no particles exist rather than an infinite sea of particles. Nevertheless, QFT does not eliminate all the difficulties raised by the Dirac sea, particularly the vacuum state possessing infinite energy, where it becomes necessary to employ renormalization to cancel the infinities that arise in the theory (´Alvarez-Gaum´e (2010)).

By the 1930s, world developments impacted the march toward a unified theory of physics. A worldwide economic downturn slowed progress. Universities found it increasingly difficult to fund research projects. Researchers were forced to seek alternative ways of making ends meet. Collaboration plummeted since sabbaticals were severely limited, and journals found it challenging to publish on the same regular basis they once had. Moreover, the 3rd Reich came to power in Germany as World War II approached. Much of the previous theoretical work in physics had been accomplished by German physicists, many of whom were Jewish. To escape the horrors of the 3rd Reich, many Jewish physicists, including Einstein, left Germany and migrated to the United States. Once the war broke out, most physicists were recruited by their respective governments to help with the war effort. The influx of talented German physicists greatly benefited the United States. During the war, significant time was devoted to studying nuclear decays primarily associated with atomic explosions.

More importantly, once the war ended, physics-related activities enjoyed an influx of ready cash, particularly in the United States, as a reward for helping with the war effort. With new resources, physicists constructed research facilities. They conceptualized and built sophisticated equipment like cyclotrons and atom smashers. The result was a boon to experimental physics as creatively designed experiments led to an explosion in new particle discoveries. The new particles displayed a myriad of diverse characteristics. The avalanche of discoveries left theorists overwhelmed. The theorists had little choice but to gather as much information as possible, then hope that advancements, built upon earlier theoretical work, could be realized.

### 2.2.6 Atomic Nuclei

The Dirac theory and the Pauli exclusion principle described atoms primarily in terms of how electrons behaved around the nucleus. At this point, much less was known about nuclei. By the mid-1960s, serious problems remained in understanding the properties of nuclei.

First, there was the problem of size. Atoms of a particular element were not always the same size. With only two particles, the proton and electron, known at the time, the hypothesis was that some protons within the nucleus contained an electron, rendering the proton electrically neutral. Experiments in beta decay showed atoms spontaneously emitted electrons from the nucleus. But the amount of nuclear electron energy far exceeded the observed electron energies in beta decay. In addition, the hypothesized spin configurations that included nuclear electrons did not square with what was observed. Eventually, the "nuclear electron" hypothesis was abandoned (Beiser (1967), § 12.2).

Two German physicists, Bothe and Becker, bombarded beryllium with alpha particles from a sample of polonium and found that the radiation emitted carried no charge. They assumed the radiation emissions were gamma rays. But to produce the results that Bothe and Becker observed would have required gamma rays with energies of about 53 $Mev$, far exceeding any observed gamma radiation energies known at the time. Finally, in 1932, Chadwick discovered the neutron. He assumed that the radiation Bothe and Becker had observed consisted of neutral particles whose mass is approximately equal to the proton. This hypothesis accounted nicely for the observed proton recoil energies. Maximum proton energy of 5.3 $Mev$ implies a 5.3 $Mev$ neutron energy instead of the 53 $Mev$ required for gamma radiation to produce the same effect. Immediately upon its discovery, the neutron was recognized as the missing ingredient in describing atomic nuclei. Its mass of slightly more than the proton, electrical neutrality, and spin of $1/2$ all fit with the observed properties of the nuclei when it is assumed that nuclei are made up solely of protons and neutrons (Beiser (1967), § 12.3).

The second mystery surrounding nuclei was the question of stability. The nucleus was conceived as a tightly packed conglomeration of positively charged protons and neutrons. Since positive electrical charges repel, how did the nucleus stay together? Many theories were advanced, but Yukawa offered the most intriguing in 1935. Rather than a static compression of unchangeable particles, Yukawa conceived the nucleus as a teeming pot of activity, where protons were transformed into neutrons and vice versa. Yukawa theorized that protons and neutrons attract one another by exchanging mesons. In 1947, Cecil Powell and his colleagues observed charged $\pi$-mesons confirming Yukawa's prediction (Beiser (1967), § 12.7).

According to the meson theory of nuclear forces, all nucleons consist of identical cores surrounded by a "cloud" of one or more mesons. Mesons can have "zero," "+," or "−" electric charges. The sole difference between a proton and a neutron was their respective meson cloud composition. Nuclear forces were created by the exchange of mesons between nucleons. The force between two neutrons or two protons was a result of exchanging a neutral meson ($\pi^0$). The force between a proton and a neutron was the result of exchanging charged mesons ($\pi^+, \pi^-$). A neutron that emits a $\pi^-$ is converted into a proton ($n \rightarrow p + \pi^-$), while the absorption of a $\pi^-$ by the proton the neutron is interacting with converts it into a neutron ($p + \pi^- \rightarrow n$) (Beiser (1967), § 12.7).

Yukawa was able to show that the exchange of mesons between nucleons led to mutually attractive forces. But if nucleons constantly exchanged mesons, why were neutrons and protons never found with other than their usual masses? The emission of a meson by a nucleon, which does not change mass, violates the conservation of energy law. Yukawa explained that the non-conservation of energy would be permissible so long as the exchange of mesons occurred within the time limits dictated by the uncertainty principle ($\Delta E \Delta T \geq h$) (Beiser (1967), § 12.7). With this information, Yukawa could predict the mass of the $\pi$-mesons. His calculations were confirmed experimentally.

The process of nuclear decay comes in three broad categories known by the names "alpha," "beta," and "gamma." All nuclear decay processes transform atoms into a more stable state, where decay becomes less likely to occur over time.

Nuclei containing 210 or more protons and neutrons become so large that the short-range forces holding the nucleus together can barely counterbalance the protons' mutual electrostatic repulsion. At that point, Alpha particles (identical to helium nuclei) are emitted from the nucleus. Alpha particles are the primary decay particles because it takes the least energy to liberate them than other combinations of particles. Alpha decay involves transforming an atom into one with two fewer neutrons and two fewer protons (Beiser (1967), § 13.3).

In 1928, Gamow modeled alpha decay by imagining that the nucleus acted like a well, where the alpha particle became trapped. Constantly jiggling, the alpha particle bounces against the well-barrier. From a classical standpoint, the alpha particle would not have enough energy to escape the barrier, so alpha decay should not happen. But Gamow used quantum mechanical principles, where an alpha particle's potential energy is averaged. He reasoned that if some of the particle collisions against the barrier carried less and some more energy than the average energy, there would be a tiny probability that an alpha particle in one of its collisions against the barrier would carry enough energy to escape. Gamow (1928) derived a relationship between the half-life of alpha decay and the energy of the alpha particle emissions previously discovered experimentally by Geiger and Nuttall. His explanation of alpha decay is one of the most striking confirmations of the correctness of quantum mechanics over classical physics.

Beta decay, like alpha decay, is a means by which atoms gain greater stability. Why beta decay occurs is still somewhat mysterious but appears related to the ratio of the number of neutrons to the number of protons. If that ratio grows much above one, the probability that an atom will undergo beta decay increases.

When first observed, beta decay displayed troubling characteristics. For instance, during decay, an electron is emitted from the nucleus.

But the nucleus was not supposed to contain electrons. The problem was overcome by what amounts to a sleight of hand by assuming that the electron left the nucleus immediately after its creation. Far more troubling, beta decay violated the conservation laws of energy, momentum, and angular momentum (Beiser (1967), § 13.5).

Beta decay involves the conversion of a neutron into a proton, represented by $n \rightarrow p + e^-$. In 1930, Pauli suggested that if an uncharged spin-1/2 particle of small or zero mass is emitted, along with the electron, then the violation of the conservation laws disappears. The hypothesized particle resolved the discrepancy between the electron's energy and the recoiling daughter nucleus, so the conservation laws held. The two hypothesized particles were the "neutrino" and "antineutrino," designated by the symbols, $\nu$ and $\bar{\nu}$. This hypothesis predicted two types of beta decay: "Positive and negative." The formulas that included the new particles could be written, $n \rightarrow p + e^- + \bar{\nu}$, $p \rightarrow n + e^+ + \nu$. The first formula signifies negative beta decay, where a neutron decays into a proton, an electron, and an antineutrino. The other is positive beta decay, where a proton decays into a neutron, a positron, and a neutrino. The neutrino was not discovered until 1959. Its discovery was difficult. Because of its electrical neutrality, it does not interact with charged particles. It also possesses either a very small or no mass at all. So feeble is its interaction with matter, it must pass through about 130 light-years of solid iron on average before it interacts with other matter (Beiser (1967), § 13.5).

The standard model of particle physics treats neutrinos as particles without mass. But more recent experiments have indicated that neutrinos have a small mass. Neutrinos have become an important topic because a neutrino with a mass suggests new physics beyond the standard model. Neutrinos are believed to be produced in great abundance in the Universe's early stages. Their tiny mass but huge numbers are thought to contribute to the total mass of the Universe and might suggest why the Universe is expanding (Fukuda (1998), pp. 1158–1162).

Like whole atoms, the nucleus can be in an excited state. However, excited nuclei return to their lowest energy state by emitting several photons corresponding to the energy differences between the excited and lowest energy states. Such emitted photons are called "gamma rays," and the process is called "gamma decay." Gamma rays are produced after the occurrence of other types of decay. The mechanism involves the nucleus emitting an $\alpha$ or $\beta$ particle so that the daughter nucleus is usually left in an excited state. It can then move to a lower energy state by emitting a gamma ray in the same way that an electron can jump to a lower energy state by emitting a photon (Beiser (1967), § 13.9).

By the 1930s, alpha, beta, and gamma decay had all been observed experimentally and explained individually. Still, no comprehensive theory of radioactive decay was formulated until the development of the electro-weak theory in the 1970s.

### 2.2.7 Toward a Theory of Elementary Particles

By the mid-1960s, about thirty varieties of relatively stable and many unstable particles had been discovered. "Relatively stable" meant particles could be detected in a bubble chamber, whereas unstable particles were nearly impossible to observe. Therefore, in an attempt to find an all-encompassing theory, the particles were organized and classified in various ways. One such organized display appears in Table 2.2.6-1[2].

---

[2] Reproduced from A. Beiser (1963), *Concepts of Modern Physics*, McGraw-Hill Inc., NY, P-378.

Table 2.2.6-1.
Elementary particles stable against decay by the strong interaction.

| | | Particle | | | Antiparticle | | | | | | | | | | | | |
|---|---|---|---|---|---|---|---|---|---|---|---|---|---|---|---|---|---|
| Class | Name | +e | 0 | -e | +e | 0 | -e | Spin | Rest Mass $m_e$ | Rest Mass Mev | Half Life Sec | L | M | B | S | Y | I |
| Photon | Photon | | $\gamma$ | | | $\gamma$ | | 1 | 0 | 0 | Stable | 0 | 0 | 0 | | | |
| Lepton | e-neutrino | | $\nu_e$ | | | $\bar{\nu}_e$ | | 1/2 | 0 | 0 | Stable | +1 | 0 | 0 | | | |
| | μ-neutrino | | $\nu_\mu$ | | | $\bar{\nu}_\mu$ | | 1/2 | 0 | 0 | Stable | 0 | +1 | 0 | | | |
| | Electron | | | $e$ | $e^+$ | | | 1/2 | 1 | .51 | Stable | +1 | 0 | 0 | | | |
| | μ-meson | | | $\mu$ | $\mu^+$ | | | 1/2 | 207 | 106 | 1.5X10 [illegible] | 0 | +1 | 0 | | | |
| Meson | π-meson | | $\pi^0$ | | | $\pi^0$ | | 0 | 264 | 135 | 7X10 [illegible] | 0 | 0 | 0 | 0 | 0 | 1 |
| | | $\pi^+$ | | | | | $\pi$ | 0 | 273 | 140 | 1.8X10 [illegible] | 0 | 0 | 0 | 0 | 0 | 1 |
| | K-meson | $K^+$ | | | | | $K$ | 0 | 966 | 494 | 8X10 [illegible] | 0 | 0 | 0 | +1 | +1 | 1/2 |
| | | | $K^0$ | | | $\bar{K}^0$ | | 0 | 974 | 498 | 7X10 [illegible] 4X10 [illegible] | 0 | 0 | 0 | +1 | +1 | 1/2 |
| | η-meson | | $\eta^0$ | | | $\eta^0$ | | 0 | 1037 | 548 | ~10 [illegible] | 0 | 0 | 0 | 0 | 0 | 0 |
| Baryon | Nucleon | $p$ | | | | | $\bar{p}$ | 1/2 | 1836 | 938 | Stable | 0 | 0 | +1 | 0 | +1 | 1/2 |
| | | | $n$ | | | $\bar{n}$ | | 1/2 | 1839 | 940 | 7X10 [illegible] | 0 | 0 | +1 | 0 | -1 | 1/2 |
| | Λ-hyperon | | $\Lambda^0$ | | | $\bar{\Lambda}^0$ | | 1/2 | 2182 | 1115 | 1.7X10 [illegible] | 0 | 0 | +1 | -1 | 0 | 0 |
| | Σ-hyperon | $\Sigma^+$ | | | | | $\bar{\Sigma}^-$ | 1/2 | 2328 | 1192 | .6X10 [illegible] | 0 | 0 | +1 | -1 | 0 | 1 |
| | | | $\Sigma^0$ | | | $\bar{\Sigma}^0$ | | 1/2 | 2332 | 1194 | <10 [illegible] | 0 | 0 | +1 | -1 | 0 | 1 |
| | | | | $\Sigma$ | $\bar{\Sigma}^+$ | | | 1/2 | 2341 | 1197 | 1.2X10 [illegible] | 0 | 0 | +1 | -1 | 0 | 1 |
| | Ξ-hyperon | | $\Xi^0$ | | | $\bar{\Xi}^0$ | | 1/2 | 2571 | 1310 | 10 [illegible] | 0 | 0 | +1 | -2 | -1 | 1/2 |
| | | | | $\Xi$ | $\bar{\Xi}^+$ | | | 1/2 | 2583 | 1320 | .9X10 [illegible] | 0 | 0 | +1 | -2 | -1 | 1/2 |
| | Ω-hyperon | | | $\Omega$ | $\bar{\Omega}^+$ | | | 3/2 | 3290 | 1676 | ~10 [illegible] | 0 | 0 | +1 | -3 | -2 | 0 |

While there are alternative ways of organizing the data, the information in Table 2.2.6-1 shows the difficulty in formulating a theory of elementary particles from an assortment of disparate data. All the particles have an antiparticle except for the photon ($\gamma$) and the $\pi^0$-meson. When a particle interacts with its antiparticle, the result is usually the complete annihilation of both, but this is not always the case. For example, if neutrons or protons interact with their antiparticles, $\pi$-mesons are usually produced, which provides evidence that $\pi$-mesons might be a mediator of the nuclear force.

Before 1956, the neutrino and antineutrino were considered the same particle. Neutrinos carried no electric charge, were massless, and differed only in spin direction. At the time, whether physical processes were right-handed or left-handed was considered immaterial. This "law of Leibniz" was a universally accepted principle called "spatial parity." It didn't matter whether a right-handed or left-handed coordinate system was used to describe a physical law, it was the same law. The neutrino and antineutrino seemed mirror images of one another, and hence, identical particles. The law of Leibniz held for nuclear and electromagnetic phenomena. But Lee and Yang noticed that many serious theoretical discrepancies could be removed if neutrinos and antineutrinos had different handedness, even though that meant they could not be reflected in a mirror. Lee's and Yang's conjectures were later experimentally confirmed. Lack of right-left symmetry in neutrinos can only occur if the neutrino is massless (Beiser (1967), § 15.2). In the standard model, neutrinos are treated as massless particles. However, current experiments have shown that the neutrinos have small masses. The issue of neutrino mass remains unresolved and suggests new physics beyond the standard model.

The $\pi$-mesons were believed to be the mediators of the nuclear force. But free $\pi$-mesons were discovered in the diffuse streams of cosmic radiation that bombard the Earth (Beiser (1967), § 15.3). Where did free $\pi$-mesons come from? A current conjecture is that they must have been abundantly produced during the big bang. In addition, charged $\pi$-mesons inevitably decay into lighter $\mu$-mesons and neutrinos. Remarkably, the neutrinos found in $\pi$-meson decay are not the same as those found in beta decay. The existence of two classes of neutrinos was established in 1962.

The discovery of the $\mu$-meson is still puzzling. Its characteristics are well known: Rest mass, 207, spin-1/2, half-life $2.26 \times 10^{-6}$ *sec*. The $\mu^+$-meson decays into a positron, a neutrino and an antineutrino. The $\mu^-$-meson decays into an electron, a neutrino and an antineutrino. The mysterious part of the $\mu$-meson is that it does not seem to have a purpose. It interacts only electrostatically with matter. Except for its mass and half-life, it is just a heavy electron. Strangely, about 99% of the time, $\pi$-mesons decay into $\mu$-mesons, and only 1% of the time, directly into electrons. Since $\mu$-mesons decay into electrons and positrons, why is there this intermediate step?

Finally, the $K$-mesons decay into $\pi$-mesons, $\mu$-mesons, electrons/positions, and neutrinos in various mixtures. They are capable of nuclear and electromagnetic interactions but interact weakly with nuclei (Beiser (1967), § 15.5).

The final class of particles is the "baryons," consisting of neutrons, protons, and hyperons. Hyperons are particles heavier than the proton and come in four subclasses in order of increasing mass. All are relatively fleeting, with extremely short half-lives. Their characteristics can be found in Table 2.2.6-1. All hyperons have definite interactions with nuclei.

An underlying structure eventually emerged from what seemed like the chaos of a vast number of particles with diverse characteristics. The structure itself did not constitute a theory of elementary particles but provided hope that a complete theoretical picture would emerge (Beiser (1967), § 15.7). Ultimately the structure became part of the standard model of particle physics.

By this time, the belief that particle interactions were mediated through an exchange of particles was well established. The photon was in a class by itself as the electromagnetic interaction mediator. But the other three force-carrying particles that would ultimately become part of the standard model ($W$, $Z$, $g$ gauge bosons) had yet to be hypothesized or discovered. Interestingly, speculation arose that a massless, stable spin-2 particle, called the "graviton," was the mediator of the gravitational force. But the graviton failed to make it into the standard model and has yet to be discovered.

Suspicions arose that the $\pi$-mesons were the mediators of the nuclear force, but nuclear interactions were the least understood at the time and, unlike the photon, $\pi$-mesons carried mass. At this point, mesons were not classified as force carriers. It was well-known that if energy conservation was preserved, the uncertainty principle required that the range of a force be inversely proportional to the mass of the exchanged particles. The electromagnetic and gravitational forces were thought to have an infinite range; therefore, their interchanging particles had zero mass (Beiser (1967), § 15.7).

The next class of particles is the "leptons" or "small particles," all with a spin of 1/2. Table 2.2.6-1 shows how they appear in the standard model, less the tau-neutrino and tau-particle. Then there are the mesons with various masses and zero spin. The final class, the baryons, included the nucleons (protons and neutrons) and the hyperons. Hyperons are heavy particles, heavier than the proton. All baryons have a spin of 1/2. Eventually, "quarks" would replace the mesons and baryons in the standard model. But, at this point, quarks had yet to be discovered.

While grouped mainly based on mass and spin, additional evidence argued in favor of classifying particles as organized in Table 2.2.6-1. Suppose three new quantum numbers are introduced; call them "$L$," "$M$," and "$B$"; assign $L = 1$ to the electron and $e$-neutrino and $L = -1$ to their antiparticles. All other particles have $L = 0$. Additionally, assign $M = 1$ to the $\mu$-meson and its neutrino and $M = -1$ to their antiparticles; all other particles have $M = 0$. Finally, assign $B = 1$ to all the baryons and $B = -1$ to their antiparticles; all other particles have $B = 0$. In every process undertaken by an elementary particle, the total values of the new $L, M,$ and $B$ quantum numbers remain constant. The classical conservation laws of energy, momentum, angular momentum, and electric charge, plus the conservation of the new $L, M,$ and $B$ quantum numbers, implied that only certain transformations among elementary particles were possible (Beiser (1967), § 15.7). To a large extent, these conservation laws were Nature's way of determining which processes could and could not occur.

Introducing the new quantum numbers, $L, M,$ and $B$, did not cover all circumstances. It was unclear why some heavy particles decayed into lighter ones with the emission of a gamma ray while others, seemingly equally permissible decays, did not occur. For instance, the $\Sigma^0$ baryon decayed into a $\Lambda^0$ baryon and a gamma ray, but the $\Sigma^+$ baryon never decayed into a proton and a gamma ray. Moreover, these heavier, soon-to-be-named "strange particles" were never created singly but always came in combinations of two or more at a time. Explaining this oddity required a new quantum number called "strangeness" ($S$). In all electromagnetic and strong interactions, $S$ is conserved. Its introduction explained why particular baryon decays occurred while others did not. Only in weak interactions did $S$ change, but only by +1 or −1 in any given decay. It was also convenient to introduce a quantity called the "hypercharge" ($Y$) ($Y = S + B$). $Y$ was conserved in all strong interactions (Beiser (1967), § 15.8).

Table 2.2.6-1 shows several particle families whose members have the same mass and interaction properties but different charges. These families can be considered single entities but in a different charge state. One way of categorizing them is by assigning the family a number, $I$, where the number of charge states is given by the formula, $2I + 1$. Hence, for the nucleon, $I = 1/2$, so that $2I + 1 = 2$. The two states are the proton and the neutron. The $\pi$-meson family has $I = 1$, so $2I + 1 = 3$. The three charge states are the $\pi^+$, $\pi^0$ and $\pi^-$. The number, $I$, was named "isotropic spin," although it has nothing to do with the spin state of a particle. Isotropic spin is analogous to quantum angular momentum, and hence, can be given a number, $I_3$, restricted to the values, $I, I-1, \ldots, 0, \ldots, -(I-1), -I$. So $I_3$ takes on half-integer values if $I$ is a half-integer and integer values if $I$ is an integer. For the nucleon, $I = 1/2$, and hence, $I_3 = 1/2$ or $-1/2$; the former is taken to represent the proton and the latter the neutron. In the case of $\pi$-mesons, $I = 1$, so $I_3 = 1$ corresponds to $\pi^+$, $I_3 = 0$ to $\pi^0$, and $I_3 = -1$ to $\pi^-$.

The charge of a meson or baryon is related to $Y$ and $I_3$ by the formula

$$q = e\left(I_3 + \frac{Y}{2}\right) = e\left(I_3 + \frac{S+B}{2}\right).$$

Hence, in the case of the nucleons, the proton has $I_3 = 1/2$, $B = 1$ and $S = 0$, so $q = e$, the opposite of the charge on an electron ($-e$). For the neutron, $I_3 = -1/2$, $B = 1$ and $S = 0$, so $q = 0$. The other meson and baryon charge states can be calculated similarly. Charge and the baryon number, $B$, are conserved in all interactions. Thus, $I_3$ is conserved whenever $S$ is, namely, in strong and electromagnetic, but not in weak interactions (Beiser (1967), § 15.9).

Such was the systematics uncovered through studying the behavior of the assortment of particles discovered through a myriad of carefully designed experiments. Yet, despite the apparent regularities and the identification of an abundance of conservation laws, the systematics did not amount to a theory of elementary particles.

# Chapter 3: The Standard Model and Beyond

***"It is often stated that of all the theories proposed in this century, the silliest is quantum theory. In fact, some say that the only thing that quantum theory has going for it is that it is unquestionably correct."***

***— Michio Kaku, Hyperspace***

## 3.0 Introduction

In Chapter Two, we learned that elementary particles were subject to several conservation laws that suggested there was a "yet-to-be-fully-understood" structure that governed all particle interactions. The critical factor in developing a theory of elementary particles was the relationship between the mathematical concept of "symmetry" and conservation laws. For example, Noether's theorem states that there is an associated conservation law for every continuous symmetry (Noether (1918)). By the mid-1960s, several symmetry/conservation relationships were well understood (see Table 3-1).

Without conservation laws, tracking the endless number of changes within each system would be virtually impossible. By concentrating on what remains unchanged during the various particle interactions, many of the variables that make up complex systems can be ignored. The connection between symmetry and conservation laws ultimately led to a theory of elementary particles.

Table 3-1[3]
The symmetries associated with particle interactions.

| Symmetry Operation | Conserved Quantity |
|---|---|
| **All interactions are independent of:** | |
| Translation in Space | Linear Momentum |
| Translation in Time | Energy |
| Rotation in Space | Angular Momentum |
| Electromagnetic Gauge Transformation | Electric Charge |
| Lorentz Transformation | Velocity of Center of Mass |
| Interchange of Identical Particles | SPIN Statistics |
| Inversion of Space, Time, and Charge | CPT Parity |
| ? | Baryon Number (*B*) |
| ? | Lepton Number (*L*) |
| ? | Meson Number (*M*) |
| **The strong and electromagnetic interactions are independent of:** | |
| Inversion of Space | Parity |
| Reflection of Charge | Charge parity, $I_3$, Strangeness (*S*) |
| **The strong interaction is independent of:** | |
| Charge | Isotropic spin (*I*) |

## 3.1 Group Theory

Table 3-1 gives an overview of the symmetries associated with particle interactions. Another way of looking at symmetry is from a mathematical standpoint. Mathematicians call the study of symmetry "group theory." A group is an abstract algebraic system, and it was unclear whether the mathematical abstractions embodied in group theory would have any connection to the physical world. Even Einstein did not recognize that his special and general theories of relativity were studies in symmetry. But, in 1931, Weyl (1950) realized that group theory could be applied to quantum physics.

Algebraically, a group is a set of elements, $G = \{a, b, c, \ldots\}$, together with an operation or binary composition rule signified by "$\circ$." The elements, in conjunction with the composition rule, create an algebraic system with the following properties:

[3] Reproduced from A. Beiser (1963), *Concepts of Modern Physics*, McGraw-Hill Inc., NY, P-386.

1. **Closure**: $a, b \in G \rightarrow a \circ b \in G$. The composition, $a \circ b$, of any two elements in $G$, is itself an element of $G$.

2. **Associativity**: $a, b, c \in G \rightarrow (a \circ b) \circ c = a \circ (b \circ c)$.

3. **Identity**: There exists a unique element, $e \in G$, such that $a \in G \rightarrow a \circ e = e \circ a = a$.

4. **Inverse**: If $a \in G$, there is a unique element, $a^{-1} \in G$, such that $a \circ a^{-1} = a^{-1} \circ a = e$.

The closure property ensures that the binary composition rule, $\circ$, does not generate elements outside $G$. Associativity makes the composition rule unambiguous. The composition, $a \circ b \circ c$, is unambiguous because the two interpretations allowed by the existence of a binary composition rule, $(a \circ b) \circ c$ and $a \circ (b \circ c)$, are equal (Vvedensky (2001), § 2.1).

Although the composition rule, $\circ$, could correspond to a numerical operation like multiplication or addition, this is not required. The composition rule simply combines a pair of group elements to obtain a third group element. Whereas ordinary numbers commute (i.e., $a \circ b = b \circ a$), the composition rule need not be commutative. In other words, $a \circ b$ need not equal $b \circ a$.

Example: Does the set of integers, $Z = \{ \ldots, -3, -2, -1, 0, 1, 2, 3, \ldots \}$, with the composition rule of ordinary addition, form a group? (i) The sum of any two integers is an integer, ensuring closure; (ii) addition of integers is associative; (iii) "0" is the identity element since $n + 0 = n, n \in Z$; (iv) the inverse of $n$ is $-n$ since $n + (-n) = 0$, and $-n$ is an integer. Thus, all four group conditions are satisfied. Hence, the integers form a group under addition (Vvedensky (2001), § 2.1).

Example: Does the set of integers, $Z$, with a composition rule of ordinary multiplication, form a group? The product of any two integers is an integer (closure is satisfied), multiplication is associative, and the unit element is 1 since $n \cdot 1 = n, n \in Z$. Still, the inverse of $n$ is $1/n$, which, in general, is not an integer. The last group condition that all the set elements must have a unique inverse is not satisfied. Hence, the integers under multiplication do not form a group (Vvedensky (2001), § 2.1).

It is not apparent that the mathematical group concept has anything to do with the symmetry of objects like people and atoms or even idealized objects like triangles or circles. An object is "symmetric" if a change in the object does not change its appearance. For example, an equilateral triangle is indistinguishable after rotations of $2/3\,\pi$ and $4/3\,\pi$ radians around its geometric center or symmetry axis. A square is indistinguishable after rotations of $1/2\,\pi$, $\pi$, $3/2\,\pi$ and $2\pi$ radians, and the circle is indistinguishable after all rotations around its symmetry axis. Such changes are called "symmetry transformations," elucidated by saying that the object remains "invariant" under a symmetry transformation (Vvedensky (2001), § 1.1.1).

The equilateral triangle and the square are objects with discrete symmetries (i.e., the symmetry rotation angles and reflections have only specific values). Therefore, there are only certain degrees for which the square remains invariant. However, the symmetries of a circle are continuous. Therefore, it can be rotated by any amount around its center and remain invariant. In the physical sciences, symmetry is fundamental because there are transformations that leave the laws of physics invariant. Such transformations involve changing the variables within a physical law (written in mathematical language) such that the equations describing the law retain their form when expressed in new variables (Vvedensky (2001), § 1.1.2).

To understand this better, consider a Galilean transformation, where the laws of physics are invariant, whether an experiment is conducted in a stationary laboratory or one traveling at a constant speed. For example, the physics governing a glass of wine is the same whether it is stationary or on a train moving at 200 miles per hour. To put the Galilean transformation in terms of symmetry, the laws of physics remain invariant if a transformation is made from a stationary state to one moving with a constant velocity.

Galilean symmetry was considered a law of Nature during Newton's time and long after. But later, it was discovered that Maxwell's theory of electromagnetism violated Galilean symmetry. Experiments with light showed that an experimenter could distinguish between standing still and moving at a constant speed. One plausible explanation was that light traveled through a luminous aether. The aether could have accounted for the discrepancy between Newton's and Maxwell's theories. When appropriately designed experiments failed to detect the aether, the contradictions between the two theories remained.

Einstein saved Galilean symmetry. Maxwell's theory postulated that light traveled at a finite constant velocity, not at an infinite velocity, as Newton's theory demanded. Consequently, the Galilean law did not just involve a spatial transformation (i.e., transforming from a stationary state to one moving with a constant velocity). "Time" had to be considered as well. Einstein turned the Galilean transformation from a purely spatial transformation to one involving space and time. If time was included, the laws of physics remained invariant under space-time transformations. This insight removed the contradiction between the theories of Newton and Maxwell. Instead of giving an observer's location in terms of three spatial coordinates, $(x, y, z)$, the time on each observer's watch had to be specified as well (i.e., $(x, y, z, t)$). Einstein's equation for representing this idea was

$$-x^2 - y^2 - z^2 + (ct)^2 = -x'^2 - y'^2 - z'^2 + (ct')^2.$$

The arrangement of minus and plus signs reflects the fact that nothing travels faster than light. The coordinates, $(x', y', z', t')$, usually denote a stationary observer while, $(x, y, z, t)$, represent an observer moving at a constant velocity. Still, this choice is arbitrary. Symmetry appears in the sense that $(x', y', z', t')$ can be replaced by $(x, y, z, t)$ and $(x, y, z, t)$ replaced by $(x', y', z', t')$ and the equation remains the same.

Two observers, one moving at a constant speed relative to a given clock and the other stationary to it, may not agree on the location and time of an event but will agree on the space-time interval represented by the equation above. In this sense, the space-time interval is "invariant" or "symmetric."

A permutation is an arrangement of $n$ distinct objects. The number of permutations is $n!$ (i.e., $n(n-1)(n-2)\dots 1$). For example, if there are three distinct objects, there are $3 \cdot 2 \cdot 1 = 6$ permutations or arrangements of the objects. These $n$ distinct object arrangements form a group, denoted by "$S_n$." The group composition rule, in this case, is the permutation of the objects.

Remarkably, the permutation group, $S_3$, is identical to the symmetry transformations of an equilateral triangle. To see this, the six elements of $S_3$ are listed below:

$$e = \begin{pmatrix} 1 & 2 & 3 \\ 1 & 2 & 3 \end{pmatrix} \qquad a = \begin{pmatrix} 1 & 2 & 3 \\ 2 & 1 & 3 \end{pmatrix} \qquad b = \begin{pmatrix} 1 & 2 & 3 \\ 1 & 3 & 2 \end{pmatrix}$$

$$c = \begin{pmatrix} 1 & 2 & 3 \\ 3 & 2 & 1 \end{pmatrix} \qquad d = \begin{pmatrix} 1 & 2 & 3 \\ 3 & 1 & 2 \end{pmatrix} \qquad f = \begin{pmatrix} 1 & 2 & 3 \\ 2 & 3 & 1 \end{pmatrix}.$$

In this notation, the top line represents the initial order of the objects, and the bottom line represents the effect of the permutation. The composition rule corresponds to performing successive permutations. It is done by rearranging the objects according to the first permutation and then using the second as the initial order to rearrange the objects according to the second permutation. So, in this case, the composition rule instructs that one permutation follows the other. As an example, consider the composition, $ad$, where permutation, $d$, is performed first, followed by permutation, $a$. Element, $d$, permutes the initial order in the following manner: (1 2 3) → (3 1 2). Element, $a$, then permutes (3 1 2) by putting the first object, in this case, 3, in the second position, the second object, in this case, 1, into the first position, and leaves the third object in position, 3 (i.e., (3 1 2) → (1 3 2)). Hence,

$$ad = \begin{pmatrix} 1 & 2 & 3 \\ 2 & 1 & 3 \end{pmatrix}\begin{pmatrix} 1 & 2 & 3 \\ 3 & 1 & 2 \end{pmatrix} = \begin{pmatrix} 1 & 2 & 3 \\ 1 & 3 & 2 \end{pmatrix} = b$$

Analogously,

$$da = \begin{pmatrix} 1 & 2 & 3 \\ 3 & 1 & 2 \end{pmatrix}\begin{pmatrix} 1 & 2 & 3 \\ 2 & 1 & 3 \end{pmatrix} = \begin{pmatrix} 1 & 2 & 3 \\ 3 & 2 & 1 \end{pmatrix} = c$$

Since $ad \neq da$, unlike ordinary numbers, permutations do not commute. Groups whose composition rules commute are called "Abelian groups." Hence, $S_3$ is a "non-Abelian" group. Recall that the set, $Z$, (the set of integers under addition) formed a group. Since ordinary numbers commute under addition, the group, $(Z, +)$, is an Abelian group.

To show that $S_3$ is, in fact, a group, examine the "group product table" shown in Table 3.1-1. The Table is somewhat misnamed since it has nothing to do with the arithmetical product of two numbers but simply displays all the possible permutations of $S_3$.

Table 3.1-1.
Permutations of $S_3$.

| | $e = 123$ | $a = 213$ | $b = 132$ | $c = 321$ | $d = 312$ | $f = 231$ |
|---|---|---|---|---|---|---|
| $e = 123$ | 123 | 213 | 132 | 321 | 312 | 231 |
| $a = 213$ | 213 | 123 | 231 | 312 | 321 | 132 |
| $b = 132$ | 132 | 312 | 123 | 231 | 213 | 321 |
| $c = 321$ | 321 | 231 | 312 | 123 | 132 | 213 |
| $d = 312$ | 312 | 132 | 321 | 213 | 123 | 231 |
| $f = 231$ | 231 | 321 | 213 | 132 | 312 | 123 |

For instance, the Table shows how to find the effect of applying the permutation, $b$, first and then $a$ (i.e., $ab$ is found by finding $b$ in the first column and then looking down that row until reaching the $a$ column). The result is $abe = ab = 312 = d$. If $a$ acts first, then $b$, the result is $bae = ba = 231 = f$. Any composition of the elements results in an element of $S_3$, satisfying the closure property of a group. Although tedious, associativity can be demonstrated. The identity element is $e$, confirmed by looking along the first row and down the first column. In this particular group, each element is its own inverse:

$aa = e$, $bb = e$, and so on, which can be verified by looking left-to-right diagonally.

A geometric realization of $S_3$ is established by considering the symmetry transformations of an equilateral triangle (Figure 3.1-2). The elements, $a$, $b$, $c$, correspond to reflections through lines that intersect the vertices at locations, 3, 1, and 2, respectively, and $d$ and $f$ correspond to clockwise rotations of $2/3\,\pi$ and $4/3\,\pi$ radians. The effect of the transformations on the vertices of the triangle is identical to the corresponding elements of $S_3$. For the equilateral triangle, the composition rule, $ad$, corresponds to a rotation followed by a reflection.

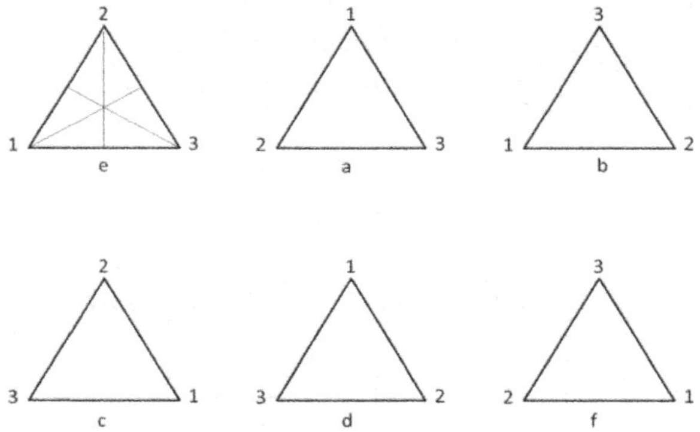


Symmetry transformations of an equilateral triangle.
Figure 3.1-2.

Thus, beginning with the standard order for the identity, the successive application of these transformations is shown below:

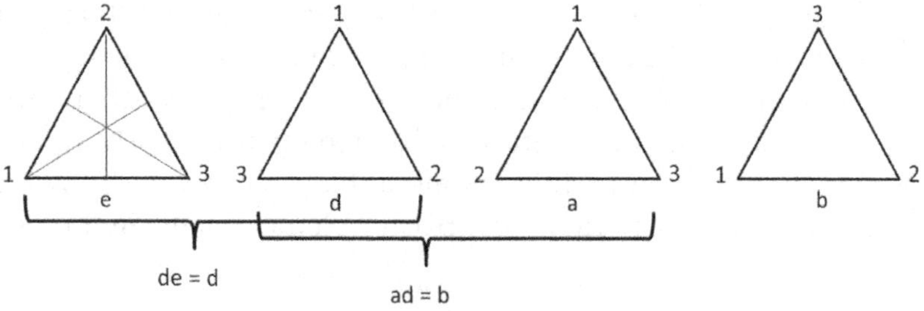

First, $d$ acts on $e$, a rotation on the original triangle. Then, $a$ acts on $d$, a reflection resulting in $b$. Similarly, $da = c$. All the compositions in $S_3$ are identical to the symmetry transformations of the equilateral triangle. If two groups have the same algebraic structure, they are identical for all intent and purposes (Vvedensky (2001), § 2.2).

There are two main points. First, the group concept, where an element acts on another element to produce a third element of the group, is very similar to how processes are described within quantum mechanics, particularly QFT. Further along, QFT will employ creation and annihilation operators that act on particles to create other arrangements of particles. Again, the mathematical way of describing this is similar to group operations.

Seemingly, group theory is a purely abstract mathematical structure with no evident connection to the physical world. However, a little less abstractly, group theory can describe the permutations of a set of objects. Even less abstractly, group theory can describe the symmetry of the equilateral triangle. The next step in the progression is to see if group theory can be applied to objects in the real world, like atoms and molecules. QFT will answer this question in the affirmative.

This last point is essential. Taking an abstract structure, a pure creation of the human mind, then applying it to the real world allows, in some sense, a human understanding of how the world works. An abstract structure is, at least potentially, understandable since the human mind created it. Finding it impossible to put our ideas into one-to-one correspondence with the structure of the world would preclude us from understanding it. The laws of Nature, whatever they might be, would be beyond human comprehension, which, in fact, they might be. Such an understanding is, to a large extent, the crux of the problem in physics today, an inability, at least so far, to combine two competing theories into one.

The theory of relativity makes mathematical sense. It is based on ordinary numbers embodied in differential calculus, which has a plausible mathematical foundation. So it seems reasonable to claim that the differential calculus is understandable to us. The quantum theory, however, is seemingly based on an entirely different set of rules. In fact, QFT is based on the Yang-Mills theory. And Yang-Mills' theory rests on shaky mathematical grounds.

### 3.1.1 The Symmetries of Nature

What kinds of symmetries are found in Nature? One of the simplest is a translation in space. Systems are generally described in terms of a set of spatial coordinates. An event takes place at a specific location described by a set of coordinates, say, $(x, y, z)$, where location is measured from a point called the "origin," commonly designated by (0,0,0). The symmetry of translation in space states that the laws of physics are independent of where the origin is chosen in space. This symmetry leads directly to the conservation of linear momentum. In QFT, the number of particles is not conserved in particle interactions. There may be more or fewer particles after an interaction than before it. But momentum is conserved. Particles, before an interaction, will have the same total momentum as after the interaction, although the particles may inhabit different positions in space.

The time translation symmetry leads to the conservation of energy. It doesn't matter what time an experiment starts. The amount of energy at the beginning of the experiment will equal the amount at the end. This symmetry is a crucial feature of Einstein's general relativity, where energy can flow with time but is never created or destroyed.

Invariance under rotations in space means that the laws of physics are independent of the coordinate system's orientation. This symmetry leads to the conservation of angular momentum (Beiser (1967), § 15.10).

Electromagnetism is governed by a "gauge symmetry." Gauge symmetry refers to shifts in the electric and magnetic field potentials. The potentials describe the electric and magnetic fields around an electromagnetic source like a magnet. There are various ways of describing the potentials, but the strength of the electric and magnetic field remains invariant under different choices for the potentials. This gauge symmetry leads directly to the conservation of electric charge.

Changes in the electric or magnetic field potentials do not change the electric charge. Closely related to gauge symmetry is the isotropy of space-time discussed earlier. The laws of physics remain invariant under space-time transformations (Feynman (1964)).

In quantum mechanics, particles are described by a symmetry that preserves the nature of the wave function. This symmetry distinguishes the force-carrying particles from the matter particles. Wave functions that obey the Bose-Einstein statistics are "bosons" - the force-carrying particles. All bosons have integer spin and do not conform to the Pauli exclusion principle. If two identical bosons are exchanged, their associated wave function remains unchanged.

On the other hand, wave functions that obey the "Fermi-Dirac statistics" are called "fermions." They have a half-integer spin and obey the Pauli exclusion principle. If two identical fermions are exchanged, their associated wave function changes sign (i.e., $\psi(x,t) \rightarrow -\psi(x,t)$). The "*SPIN* statistics" theorem states that no process occurring in an isolated system can change the statistical behavior of that system. In other words, in any isolated system of elementary particles, a boson cannot be changed into a fermion and vice versa. The nature of the two types of elementary particles (fermions and bosons) is conserved in any process in Nature (Beiser (1967), § 15.10).

The symmetries associated with the conservation of the quantum numbers, $B, L, M$, discussed earlier, remained unknown during much of the 1960s. This deficiency was addressed by introducing the quark model and the electro-weak theory in the 1970s. Gell-Mann and Zweig formulated the quark model shortly after Gell-Mann's 1961 formulation of a particle classification scheme known as the "Eightfold way" – or, in more technical terms, "$SU(3)$ flavor symmetry." Ne'eman independently developed a scheme similar to the Eightfold Way in the same year (Pickering (1984), pp. 114–125, Gell-Mann (1961)).

The remaining symmetries involve parity of one kind or another. Roughly speaking, parity can be thought of as mirror image symmetry. From a physics standpoint, parity refers to the behavior of the wave function under the inversion of space. If $\psi(x,t) = \psi(-x,t)$, the wave function has even parity. If $\psi(x,t) = -\psi(-x,t)$, the wave function has odd parity. Whether even or odd, a system retains its parity throughout an interaction. As was learned earlier, $\psi(x,t)$ has no physical interpretation in quantum mechanics. Instead, $\psi^2(x,t)$ is related to an actual outcome and then only to the probability of the outcome happening. Hence, outcomes are independent of parity.

So why is parity important? Parity can limit the number of outcomes in an isolated system. It is possible to describe systems with a certain parity in an initial state and then predict interactions with equally plausible final states that do not occur in Nature. The conservation law is the conservation of parity, which states that whether described by a left-handed or right-handed coordinate system, the physics of an isolated system will remain the same (Beiser (1967), § 15.10).

Interestingly, there are particles in Nature that violate spatial parity ($P$). For example, a neutrino cannot be described using a right-handed coordinate system, nor can an anti-neutrino be described using a left-handed coordinate system. The handedness results from neutrinos partaking in the weak interaction. Spatial parity is not conserved in weak interactions but only in strong and electromagnet interactions.

There are two more essential parities: Time (T) and charge (C). Time parity is the symmetry that says $\psi(x,t)$ does not change if $t$ is replaced by $-t$. This symmetry is called "time reversal." If a process can occur in Nature, its reverse occurs. For a long time, T was considered a symmetry of Nature. But in 1964, the $K_2^0$ meson decayed into a $\pi^+$ and a $\pi^-$, which violated the conservation of T.

Charge conjugation, C, is the symmetry associated with replacing all the particles of a system with their antiparticles. Any process under the strong or electromagnetic interactions will have an identical process for its antiparticles. But C symmetry fails in weak interactions. However, if all three parities are taken together (CPT), then this is a symmetry of Nature: For every process, an antimatter mirror-image counterpart takes place in reverse. And this happens even though each parity fails individually during some interactions (Beiser (1967), § 15.10).

Physicists face two challenges in constructing a unified theory based on symmetry. Firstly, not all interactions exhibit the same symmetries. For instance, the strong interaction is independent of isotopic spin, but this is not the case with the other interactions. Secondly, Nature is not entirely symmetrical. Nature's symmetries are, in a sense, hidden. It is the invisible world of the wave function that is symmetrical. Upon observation, the symmetries disappear and leave only an asymmetrical outcome. The weak interaction involves "symmetry breaking," which explains how particles, in particular, *W* and *Z* bosons, gain mass.

## 3.2 The Standard Model of Particle Physics

The search for a theory of elementary particles culminated in the standard model of particle physics, which is not a completely predictive theory. Its predictive power requires experimental input as well as theoretical insight. The standard model is a hybrid, part experimentally derived inputs, part theory. But once the input parameters are known, the standard model makes unassailable predictions in the domain where it has been tested.

With that in mind, what constitutes the standard model? Table 3.2-1 shows the standard model's current elementary particle classification scheme. There are 17 known fundamental particles. The Higgs boson was discovered in the summer of 2012 at the European Organization for Nuclear Research (CERN).

Physicists have identified four fundamental interactions or forces that govern the behavior of elementary particles. First, the strong force is primarily responsible for binding the nucleus together. Electromagnetic forces describe interactions between charges, while weak forces primarily describe nuclear decay. Finally, the gravitational force is absent from the standard model.

Table 3.2-1.
Elementary particle classification scheme.

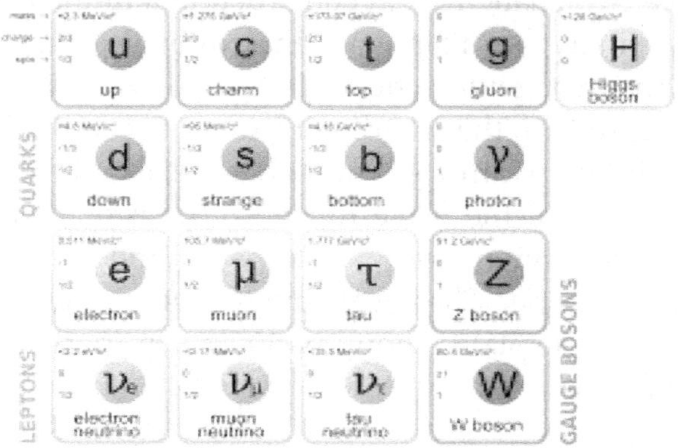


The two major elementary particle groups are the fermions and the bosons. Fermions have half-integer spin, while bosons have integer spin. Most bosons have a spin of 1, including the photon. There are $W^+$, $W^-$ and $Z^0$ bosons, the force carriers of the weak interaction, all have a spin of 1. Each elementary particle is associated with an antiparticle with the same mass but an opposite quantum number. Some particles, such as the photon, are identical to their antiparticle. Such particles must be neutral (no electric charge), but not all neutral particles are identical to their antiparticle. Particle-antiparticle pairs can annihilate each other, releasing energy equal to twice the particle's rest energy. The antiparticle is often denoted by the same symbol as the particle but with a bar over the symbol. For example, the antiparticle of the proton, $p$, is denoted by $\bar{p}$.

The fermions are subdivided into "leptons" and "quarks." Leptons do not participate in strong interactions and are freely propagating particles. Quarks come in six varieties called "flavors:" Up *(u)*, down *(d)*, charm *(c)*, strange *(s)*, top *(t)*, and bottom (sometimes referred to as "beauty") *(b)*. Quarks appear in pairs or triplets with other quarks and antiquarks. The resulting particles are the hadrons (more than 200), which include the baryons and the mesons.

A quick comparison of Table 2.2.6-1 with Table 3.2-1 reveals that the standard model's particle classification scheme contains no hadrons. In 1964, Gell-Mann and Zweig hypothesized that hadrons were composite particles composed of combinations of more elementary particles that Gell-Mann named "quarks." Like other elementary particles, each quark has an antiquark. Their model involved the three quarks. $u, d, s$ (Pickering (1984), pp. 114-125, Gell-Mann (1961)).

Initially, the theory of quarks received a muted reception. Quarks were never isolated and possessed a fractional electric charge, generating suspicion because such particles had never been observed. Moreover, it was unclear whether the quark was a fundamental particle or a convenient abstraction designed to explain concepts not fully understood at the time. But shortly, Glashow and Bjorken predicted a fourth flavor of quark, which they named "charm." It led to a better description of the weak interaction. This discovery made the number of known quarks equal to the number of known leptons and delivered a mass formula that correctly reproduced the masses of the known mesons (Bjorken (1964), pp. 255–257, Gell-Mann (1964), pp. 214–215).

The theory that describes the strong interaction is called "quantum chromodynamics" (QCD) and is governed by the $SU(3)$ symmetry group. Strong interactions remain invariant in an internal color space. There are three distinct color charges (red, blue, and green). The charges have nothing to do with color. The term is used mainly to distinguish the strong from the electromagnetic force. Regarding group theory, the assertion that there are no color singlets is simply the statement that QCD has an $SU(3)$ rather than a $U(3)$ symmetry. The experimental evidence supports the $SU(3)$ group structure. If color singlets existed, then, mathematically speaking, the elements of the group would commute, which suggests an Abelian group. $U(3)$ is an Abelian group, $SU(3)$ is not (Griffiths (1987), pp. 280–281).

In 1968 scattering experiments at the Stanford Linear Accelerator Center (SLAC) showed that the proton was made of smaller point-like objects, providing evidence that the proton was not an elementary particle. The point-like objects were later identified as the $u$ and $d$

quarks. The strange quark turned out not only to be a necessary component of Gell-Mann's and Zweig's three-quark model, but it explained the characteristics of the $K$ and $\pi$-mesons. SLAC indirectly validated the strange quark experimentally. The number of quark flavors grew to six in 1973 when Kobayashi and Maskawa discovered that CP violation could be explained if another quark pair existed (Breidenbach (1969), pp. 935–939).

In November 1974, charm quarks were produced in the lab almost simultaneously by two teams—one at SLAC under Burton Richter and the other at Brookhaven National Laboratory under Samuel Ting. Charm quarks were observed bound with charm antiquarks in mesons. This discovery convinced most physicists that quarks were fundamental particles, not just convenient abstractions. In 1977, the bottom quark was observed by a team at Fermilab led by Leon Lederman. Finally, in 1995 the top quark was also observed at Fermilab. It had a mass much greater than expected —almost as great as a gold atom. Table 3.2-2 shows the quark makeup of various baryons and mesons (Carithers (1995), pp. 4–16).

Table 3.2-2.
Quark makeup of various baryons and mesons.

| Mesons | | | Baryons | | |
|---|---|---|---|---|---|
| **Name** | **Particle** | **Quark Makeup** | **Name** | **Particle** | **Quark Makeup** |
| *π-meson* | $\pi^0$ | $u\bar{u} - d\bar{d}$ | *Nucleon* | $p$ | *uud* |
| *π-meson* | $\pi^-$ | $u\bar{d}$ | *Nucleon* | $n$ | *ddu* |
| *K-meson* | $K^+$ | $u\bar{s}$ | *Hyperon* | $\Lambda^0$ | *uds* |
| *η-meson* | $\eta^0$ | $u\bar{u} + d\bar{d} - 2s\bar{s}$ | *Hyperon* | $\Sigma^0$ | *uds* |
| | | | *Hyperon* | $\Sigma^+$ | *uus* |
| | | | *Hyperon* | $\Sigma^-$ | *dds* |
| | | | *Hyperon* | $\Xi^0$ | *uss* |
| | | | *Hyperon* | $\Xi^-$ | *dss* |
| | | | *Hyperon* | $\Omega^-$ | *sss* |

### 3.2.1 The Characteristics of the Standard Model

The standard model is much more than an elementary particle classification scheme. It is a powerful predictive tool. Given the input parameters and fundamental particle spectra, the standard model makes specific predictions for the structure of composite systems, such as nucleons, nuclei, atoms, and their interactions, such as collisions or decays. The theoretical part of the standard model is a relativistic QFT in which local gauge symmetries describe the dynamics of the field.

Each interaction is governed by a Lagrangian density function ($\mathcal{L}$), which describes the dynamics of the system and provides the equations of motion (i.e., there is an $\mathcal{L}$ for each interaction ($\mathcal{L}_{QED}$, $\mathcal{L}_{ew}$, $\mathcal{L}_{QCD}$)). Each Lagrangian density function must remain invariant under the gauge symmetry that constrains it. A gauge symmetry (also called "gauge invariance") is the property in which different configurations of the underlying fields, which are not directly observable, result in identical observable quantities. A transformation from one field configuration to another is called a "gauge transformation." Lagrangian density functions remain invariant under gauge transformations. For example, the electric and magnetic fields, $E$ and $B$, are observable, while their potentials, $V$ (voltage) and $A$ (the vector potential), are not. Under a gauge transformation, in which a constant is added to $V$, there is no observable change in $E$ or $B$ (Goshaw (2012), pp. 6-8, Klauber (2012), § 6).

The gauge symmetries constrain the possible choices for a theory governed by a particular Lagrangian. For instance, the choices for a $\mathcal{L}_{QED}$ (the Lagrangian for the electromagnetic interaction) are constrained to equations that remain invariant under the $U(1)$ symmetry group. Gauge symmetries constrain the laws of physics because all the changes induced by a gauge transformation have to cancel each other when written in terms of observable quantities (Klauber (2012), § 8).

Modern physical theories describe reality in terms of fields (e.g., electromagnetic and gravitational fields). There are fields for the electron and all other elementary particles. In QFT, particles arise through the quantization of the fields. When a field is quantized, the continuous values of the fields are converted into a smaller number of discrete quantities.

In all of the gauge transformations of the standard model, the symmetries relate only to the matter and force fields themselves and are independent of space-time transformations. This fact represents a significant impediment in the search to unite general relativity with the standard model. This is because space-time transformations are gauge transformations not observable according to quantum theory. But space-time transformations represent gravitational effects, so they must be observable. The upshot is that gravitational effects are quantum mechanically unobservable. So far, no one has found a solution to this dilemma.

The QFT associated with the standard model is an entirely renormalizable theory, meaning that divergences (sums to infinity) appearing in specific mathematical calculations can be made finite by absorbing them into the parameters, such as masses and coupling strengths (Goshaw (2012), pp. 6-8).

The standard model possesses a few shortcomings. Some calculations are complicated, limiting the predictive power of the model. Consequently, its predictions become wrong outside the range of current experiments, particularly at very high energies. But there are no violations of the standard model's predictions in the domain where it has been quantitatively tested. That may change since the model is expected to be tested at CERN in ways it never has before (Goshaw (2012), pp. 6-8).

There is no natural mechanism for electroweak symmetry breaking (explanation for why the *W* and *Z* boson masses are not zero). This mechanism is inserted "by hand" into the standard model and leads to the prediction of the Higgs boson. While the Higgs boson has been discovered, the theoretical part of the standard model does not explain how the Higgs mechanism arises (Goshaw (2012), p. 9).

### 3.2.2 Experimental Inputs to the Standard Model

Before making predictions, the standard model requires 19+ measured parameters in its current form and about 25 measured parameters if neutrinos have mass (Goshaw (2012), p. 10).[4] The structure of the theoretical part of the model constrains the number of experimentally derived inputs. QFT requires particle-antiparticle symmetry arising from CPT invariance. Every particle has an antiparticle with the same mass and mechanical spin but opposite in the other internal quantum numbers (e.g., electric charge). In counting the experimental inputs to the standard model, tally up only the particles because their antiparticles are required by QFT (Goshaw (2012), pp. 16-19).

The symmetries governing the standard model predict the number and character of the force-carrying bosons ($\gamma, W^{\pm}, Z^0$, 8 bi-colored gluons). The theory sets these particles and are not experimental inputs (Goshaw (2012), p. 20).

The QFT part of the standard model contains incomplete information about the fermions (quarks and leptons). However, there are some constraints imposed by the theory. First, quarks must come in (electroweak) doublets. A quark doublet is called a "family" or "generation." This characteristic of quarks is required by the $SU(2)$ symmetry of the theory. The quark doublets must be paired with lepton doublets, which is required to retain renormalizability. If the symmetries of the theory are to be maintained, then the sum of all the charges of quarks and leptons must be zero. The symmetry constraint requires equal quark and lepton generations (Goshaw (2012), pp. 21-22).

QFT does not constrain the number of generations of quarks and leptons. Constructing a consistent "standard model" with 1, 2, 3, 4, or more generations of quarks/leptons is possible. This number is an input to the standard model determined by Nature. For example, a "standard model" could be constructed with two quark/lepton

[4] Note: when the standard model was first completed, it was thought the neutrinos were massless. But recent experiments show this not to be case.

generations. But, this "standard model" would not accommodate the violation of CP symmetry in electroweak interactions. The violation of CP invariance requires a "standard model" with three generations of leptons and quarks. Nature may contain other fundamental particles, but current experiments have not found any (Goshaw (2012), pp. 26-29).

The values gained from experiments, which are needed to make the standard model predictive, are listed in Table 3.2.2-1.

Table 3.2.2-1.
Experimental inputs to the standard model.

| Type of Input | Symbol | Measured Values |
|---|---|---|
| **Mass:** | | |
| Leptons | $m_e, m_\mu, m_\tau$ | 0.511 MeV, 0.106 GeV, 1.78 GeV |
| Quarks | $m_u, m_d, m_s, m_c, m_b, m_t$ | 2 MeV, 5 MeV, 104 MeV 1.27 GeV, 4.20 GeV, 175 GeV |
| Bosons | $m_W, m_Z, m_H$ | 80.40 GeV, 91.19 GeV, ~126 GeV |
| **Coupling Strength:** | | |
| coupling constants | $\alpha_e, \alpha_s$ | Depends on energy level |
| **Electroweak Mixing Parameters:** | | |
| *3* angles, *1* phase | $\theta_1, \theta_2, \theta_3, \delta$ | Depends on circumstances |
| ***QCD CP* violating parameter:** | | |
| ? | $\Lambda_{QCD}$ | *~0* |

The distinguishable fundamental particles in the standard model are six quarks, each coming in three generations, totaling 18, and adding six leptons for a total of 24 leptons and quarks. Each lepton and quark has an antiparticle for a total of 48. There are 12 gauge bosons ($\gamma, W^{\pm}, Z^0$, 8 bi-colored gluons) and the Higgs for a total of 61. But, the number of input parameters required by the standard model is not so large because QFT with CPT invariance requires that $Mass(particle) = Mass(antiparticle)$, and the QCD structure requires that $Mass(q_r) = Mass(q_g) = Mass(q_b)$. In other words, the masses of each quark color must be equal. QFT requires that the photon and gluon masses be zero.

As free particles, the lepton masses can be measured directly, but quarks are bound in color singlet hadrons (except for the top quark), and their masses must be found indirectly. All neutrino masses are zero in the current standard model (Goshaw (2012), § 2).

There are two "coupling constants." One is associated with the electromagnetic interaction, the other with the strong interaction. In mathematical calculations, coupling constants relate to the strength of the interaction. However, the term "constant" is a misnomer since neither is constant. Instead, their values depend on the energy scales at which they are measured. Technically, these parameters should come directly from the theoretical part of the model, but they don't. The situation is troubling and is related to the hierarchy problem and the issue of fine-tuning, where adjusting parameters very precisely, sometimes by many orders of magnitude, is required so that the outputs of the model agree with observations. Fine-tuning is less of an issue for the electromagnetic coupling constant, but it, too, needs a slight adjustment in higher energy experiments. More problematic is the strong interaction coupling constant; its larger value requires considerably more fine-tuning.

The electromagnetic coupling constant is $\ll 1$, so in the mathematical series expansions, after a few terms, the remaining terms do not contribute much to the series and can be ignored, which is often referred to as a "perturbation." But the strong interaction coupling constant is much larger, and perturbation techniques are not as effective or, in some cases, can't be used, making calculations more difficult.

Three angles and one complex phase parameterize the electroweak mixing parameters. The phase angle, $\delta$, is the source of CP violation and is relevant to the cosmological separation of matter from antimatter. Mathematically, these parameters are used in the CKM matrix that describes the probability of a transition from one quark flavor to another (Goshaw (2012), § 2, p. 10).

In QFT, the theory's Lagrangian of the QCD sector contains a term that would permit CP violation in the strong interaction and create an electric dipole moment in the neutron. But, no CP violation has been observed in strong interactions, and the measured dipole moment of the neutron is several orders of magnitude smaller than the theory predicts. Hence, that term in the QCD Lagrangian is augmented by a measured constant, $\Lambda_{QCD}$, which is not fixed by the theory. This anomaly is another example of fine-tuning, often referred to as the "strong CP violation problem" (Goshaw (2012), § 2, p. 11). Moreover, slight variations of some input parameters, fine-tuned to our Universe, result in dramatically different predictions. For example, a slight percentage variation in the coupling strengths results in the prediction that nuclei would be unstable (Goshaw (2012), § 2, p. 12).

### 3.2.3 The Theoretical Part of the Standard Model

The theoretical part of the standard model is based on a QFT. Non-relativistic quantum mechanics (NRQM) cannot treat interactions between particles, such as a particle and antiparticle annihilating, yielding neutral particles. Moreover, there is no way of describing how particles decay into other particles. QFT provides a means whereby particles can be annihilated, created, and transmigrated from one type of particle into other(s).

The number of particles is conserved in NRQM, where the number and types of particles remain the same in all processes. QFT does not require particle number conservation, which is more in line with what is observed in Nature (Klauber (2012), § 1.2).

QFT is a relativistic theory consistent with Einstein's theory of special relativity. Extrapolation of NRQM to Dirac's relativistic quantum mechanics (RQM) results in unobservable states with negative energies. By contrast, QFT replaces the negative energy states with antiparticles with positive energies, overcoming the earlier objections to RQM (Klauber (2012), § 1.3).

Figure 3.2.3-1 illustrates the interaction between an electron and a positron. Known as "Bhabha scattering, " it is an example of a problem QFT handles well. At $x_2$ the electron and positron annihilate to produce a "virtual photon" ($\gamma$). At $x_1$, the photon is transmuted back into an electron and a positron. Antiparticles are represented by lines with arrows pointing opposite their direction of travel in time. The strange, reverse order of numbering in time (i.e., 2 → 1) is standard in QFT (Klauber (2012), § 1.3). The annihilation of the electron and the positron at $x_2$ is accompanied by the creation of a photon, a boson (force-carrying particle). Unlike electrons and positrons, it is not a "real" particle but a transitory, short-lived, and undetectable object called a "virtual particle." The photon governs the nature of the interaction between real particles. Next, at $x_1$, the photon is annihilated, and an electron and positron are created (Klauber (2012), § 1.3).

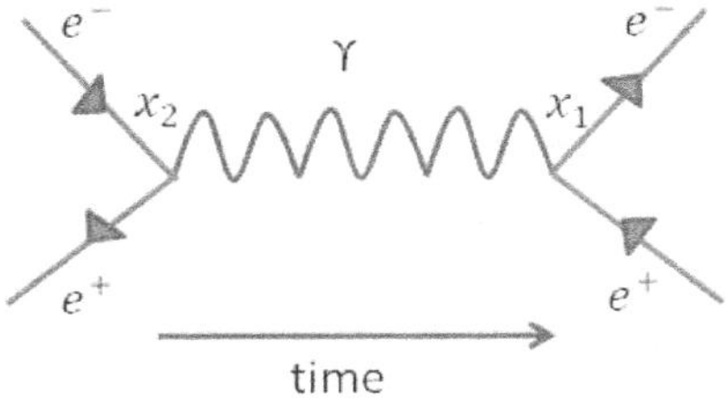


Figure 3.2.3-1.

Figure 3.2.3-1 shows a "Feynman diagram," the primary method of pictorially representing particle interactions and decays. Feynman diagrams appear simple, but each line represents a mathematical equation that can be reasonably complicated.

Mathematically, QFT delivers a "transition amplitude," which describes a transition from an initial configuration of particles to a final configuration. The square of the transition amplitude equals the probability of finding (upon measurement) that the interaction occurred. This process is similar to the NRQM case, where the square of the wave function is proportional to the probability of finding a particle at a particular location (Klauber (2012), § 1.3).

QFT employs creation and annihilation operators that act on states. States represent particle configurations, and operators represent transmutations into other particle configurations. The initial state of a quantum system, represented by the ket, "$|i\rangle$," indicates the initial configuration of particles. The final state of the system is signified by the bra, "$\langle f|$," indicating the final configuration of particles:

$$\text{Transition Amplitude} = \langle f|o_1 o_2 ... o_n|i\rangle.$$

The $o_k$'s, $k = 1,2,\ldots,n$, inserted between $|i\rangle$ and $\langle f|$, are creation and annihilation operators that act in reverse order. "$o_n$" operates on the initial state, signifying a transmutation into an altered state. The next operator acts on the altered state, and the process repeats until reaching the final state.

The process associated with QFT is similar to permutating three objects, where the initial state is represented by $|123\rangle$. When a permutation takes place, the initial state is transformed or rearranged (i.e., $a|123\rangle = |213\rangle$). The permutation, $a$, transforms $|123\rangle$ into $|213\rangle$. The permutation, $b$, transforms $|213\rangle$ into $|231\rangle$, the final state. Symbolically, this process can be written $\langle 231|ba|123\rangle$, where $|i\rangle = |123\rangle$ and $\langle f| = \langle 231|$ and $a$ and $b$ represent the operators (Sec. 3.1).

Figure 3.2.3-1 illustrates, in a grossly oversimplified manner, how a transition amplitude is obtained in QFT. First, write down the entire transition amplitude in symbolic terms:

$$\text{Transition Amplitude} = \langle e^+e^-|O_{x_1}O_{x_2}|e^+e^-\rangle.$$

The ket, $|e^+e^-\rangle$, represents the initial state, where an electron and positron are the incoming particles. At $x_2$, the transmutation is represented by the operator, $O_{x_2}$, acting on $|e^+e^-\rangle$. This operator destroys both the electron and positron and creates a virtual photon. $O_{x_2}$ can be written in a more descriptive form, $P_c\bar{\psi}_d\psi_d$. At the next event, $x_1$, the operator, $O_{x_1}$, destroys the photon and creates an electron and positron. $O_{x_1}$ can be written in a more descriptive form, $\bar{\psi}_c\psi_c P_d$. After the event at $x_1$ the system reaches its final state, represented by $\langle e^+e^-|$.

This transition amplitude can be written symbolically:

$$\text{Transition Amplitude} = \langle e^+e^- | (\bar{\psi}_c \psi_c P_d)_{x_1} (P_c \bar{\psi}_d \psi_d)_{x_2} | e^+e^- \rangle,$$

replacing the operators, $O_{x_2}$ and $O_{x_1}$, with their more descriptive forms. At the event, $x_2$, the operator, $O_{x_2}$, acts on the state, $|e^+e^-\rangle$, where $\psi_d$ destroys the electron and $\bar{\psi}_d$ destroys the positron. In other words, at the event, $x_2$, the incoming particles (ket) are destroyed by the annihilation operators, $\psi_d$ and $\bar{\psi}_d$, so at an intermediate point,

$$\text{Transition Amplitude} = \langle e^+e^- | (\bar{\psi}_c \psi_c P_d)_{x_1} (P_c)_{x_2} K_2 | 0 \rangle.$$

The annihilation of the electron and positron leaves the "vacuum ket," denoted $|0\rangle$, which signifies that no real particles remain. At this point, the vacuum ket is accompanied by a purely numeric factor, $K_2$. The formal mathematics of QFT determines the value of $K_2$ (Klauber (2012), § 1.3). Understanding what the numerical factors represent is unimportant at this stage and can be ignored without much loss of comprehension.

Next, the creation operator, $P_c$, creates a virtual photon ($\gamma$) at the event, $x_2$. The virtual particle is called a "propagator." The photon then propagates from $x_2$ to $x_1$, where the operator, $P_d$, annihilates it. This process leaves $|0\rangle$ along with an additional numeric factor, $K_\gamma$, coming from the formal mathematics:

$$\text{Transition Amplitude} = \langle e^+e^- | (\bar{\psi}_c \psi_c)_{x_1} K_\gamma K_2 | 0 \rangle.$$

At $x_1$, the remaining creation operators, $\bar{\psi}_c, \psi_c$, create an electron and positron out of the vacuum, which creates $|e^+e^-\rangle$ plus a numerical factor, $K_1$, part of the formal mathematics of QFT. The ket and the bra now represent the same state:

$$\text{Transition Amplitude} = \langle e^+e^- | K_1 K_\gamma K_2 | e^+e\text{-} \rangle.$$

Since $K_1, K_\gamma, K_2$ are numbers multiplied together and not operators, they can be removed from the bracket:

$$\text{Transition Amplitude} = K_1 K_\gamma K_2 \langle e^+e^- | e^+e^- \rangle.$$

An important point here is that the bracket of a multi-particle state is defined so that $\langle e^+e^- | e^+e^- \rangle = 1$. Hence, the

$$\text{Transition Amplitude} = K_1 K_\gamma K_2.$$

The process illustrated in Figure 3.2.3-1 is an evolution of the original state, represented by the ket, to the final state, represented by the bra. Operators act on the ket at each step to transform it into a new state. When the entire transition is complete, the bracket equals unity. What is left is the numerical quantity, $K_1 K_\gamma K_2$ (Klauber (2012), § 1.3).

The quantity, $(K_1 K_\gamma K_2)^2$, equals the probability that the process described in Figure 3.2.3-1 happens. Quantum processes depend on particle momenta, spins, rest masses, strengths of the interactions, and many other things, which all play a role in determining the probability that the interaction occurs. These, among other subtleties, have been omitted to convey, as simply as possible, the essence of a transition amplitude (Klauber (2012), § 1.3).

In NRQM, the solutions to the relevant Schrödinger equation are states (kets). But, the solutions to the relevant wave equations in QFT are not states (particles) but operators that create and annihilate particles. In this difference lies the power of QFT (Klauber (2012), § 1.3).

Each standard model sector is associated with a Lagrangian density function, which describes the system's dynamics, provides the equations of motion, and determines which particle interactions and decays can occur. In NRQM, solutions to the equations come from an appropriately chosen wave function, $\psi$, where the wave function is considered a state. By contrast, in QFT, the solutions to the equations are not states but creation and annihilation operators (things that create or destroy particles). Since a symmetry constrains each of the Lagrangian density functions associated with an interaction, the operators must be chosen so that the Lagrangian density function remains invariant under whatever symmetry governs the interaction.

The discussion here has described the theoretical part of the standard model in grossly oversimplified terms. While the details are complicated, the high-level process is relatively simple. Creation and annihilation operators determine what kinds of particle interactions are possible in Nature. As solutions, the operators must conform to certain symmetry constraints. That pretty much sums up the whole of QFT.

### 3.2.4 Summary of the Standard Model

To summarize, the standard model of particle physics describes the subatomic particles in terms of electromagnetic, weak, and strong interactions. Developed throughout the mid to late twentieth century, the standard model was molded by experimental discoveries and theoretical advances. It was a collaborative effort involving physicists from many continents, and its development took decades. Finally, the current formulation was completely experimentally validated in 2012 with the discovery of the Higgs boson. Because of its success in explaining various experimental results, the standard model is sometimes regarded as "the theory of almost everything" (Oerter (2006), p. 2).

But the standard model harbors several deficiencies:

1. The self-consistency of the theoretical part of the model has not been mathematically demonstrated.

2. Experiments indicate that neutrinos masses are not zero, which the current standard model does not allow.

3. The standard model does not include gravity, and there is no known way of describing general relativity in terms of a QFT.

4. The standard model requires 19+ unexplained numerical constants, and including the neutrino masses will require an additional 7 or 8 constants, which will likely remain unexplained.

5. The Higgs mechanism gives rise to the hierarchy problem if new physics is present at high energies, leading to fine-tuning the standard model parameters through renormalization, which is related to deficiency #1.

6. The standard model cannot explain the amount of cold dark matter and predicts dark energy contributions that are too large and cannot easily accommodate the predominance of matter over antimatter.

7. The explanation for the evolution of the Universe seems to require a mechanism such as "cosmic inflation," but no such mechanism exists in the standard model.

## 3.3 Beyond the Standard Model

The deficiencies of the standard model inspired physicists to look beyond - to find a theory that could explain the mysteries that the standard model could not. There emerged several candidates, albeit with deficiencies of their own.

### 3.3.1. Einstein's Failure

While new discoveries came to light continuously, Einstein pursued a unified field theory. While aware of the new quantum theory, he despised its indeterministic characteristics. He believed that the proper course to unification involved extending the field equations of general relativity into the quantum realm. It remained for him to show how this could be accomplished.

Einstein tried several approaches. After completing the theory of general relativity, his early work on unification was, by and large, a reaction to approaches advanced by others, including (i) the geometrization of the electromagnetic field proposed in 1918 by Weyl, (ii) the five-dimensional theory suggested by Theodor Kaluza in 1919, and (iii) the affine connection on the metric field advanced by Arthur Eddington in 1921. Einstein continued to explore modifications of the Kaluza approach until he died in 1955 (Sauer (2007), § 3).

In Weyl's geometrization of the electromagnetic field, the solution included the notion that light emitted by a radiating atom would depend on the prehistory of the atom. However, Einstein remarked that, in such a case, no uniquely defined frequencies of the spectral lines of a chemical element could exist (i.e., the frequencies would depend on the emitter's location) (Goenner (2004)), which contradicted experimental facts.

Regarding Eddington's affine connection on the metric field, Einstein found that the theory did not account for the differences in electron-proton masses and that no free electron solutions seemed possible. And since an electron is a free particle, Eddington's solution was a nonstarter (Sauer (2007), § 3).

In 1931, Einstein and his assistant Mayer derived the gravitational and electromagnetic field equations from Kaluza's five-dimensional approach. However, in their first paper, they could not account for the structure of matter and concluded that the existence of charged particles was incompatible with field equations. In the end, Einstein and his coworkers found it impossible to describe particles by non-singular solutions and to account for the quantitative difference in the strength of the gravitational and electrostatic forces between material particles (Sauer (2007), § 3).

Einstein only concerned himself with explaining how the electron and proton emerged. More serious from the conceptual point of view was the discovery of mesons. The muon was discovered in 1937. The charged pion was postulated in 1935 and discovered in 1947, and the neutral pion was postulated in 1938 and discovered in 1950. By the mid-fifties, a dozen elementary particles were known at the time of Einstein's death. Ten years later, the list of experimentally confirmed subatomic particles numbered about one hundred (Sauer (2007), § 1). With the discovery of elementary particles beyond the electron and proton, twentieth-century physics had postulated two new kinds of fundamental interactions (the weak and strong atomic interactions). Physics was moving past the older master.

### 3.3.2 Supersymmetry

Supersymmetry postulates that every elementary particle of integer spin has a "super-partner" particle of 1/2 spin. This doubling imposes a symmetry between the fermions and the bosons. For every quark (spin-1/2), there is a squark (spin-0). For a photon (spin-1), there is a photino (spin-1/2), etc.

Potentially, supersymmetry could provide the added ingredient necessary for resolving several theoretical problems. Hence, many theoretical formulations, including string theory, contain supersymmetry. For example, if supersymmetry exists at high energy scales, it could resolve the hierarchy problem. In addition, the Higgs boson mass is subject to quantum fluctuations that would make its mass so large as to undermine the internal consistency of the standard model. Currently, the problem is overcome by fine-tuning the parameters of the model. In supersymmetric theories, the contributions of the super-partners cancel quantum fluctuations more naturally, eliminating the need for fine-tuning.

Supersymmetry allows for the high-energy unification of the weak, strong, and electromagnetic interactions and provides a candidate for dark matter. The standard model suggests that the three interactions are unified at high energies. But this unification can only be accomplished with extreme fine-tuning, and even then, the forces don't entirely unify at those energies. Supersymmetry closes the gap allowing near-exact unification of the three forces without extreme fine-tuning.

The gauge symmetries of the standard model are internal symmetries, meaning they are independent of space-time transformations. The Coleman–Mandula theorem prohibits space-time and internal symmetries from being combined nontrivially. Supersymmetry appears to offer the only loophole to this. The "Haag-Lopuszanski-Sohnius theorem" demonstrates that supersymmetry is the only way space-time and the internal symmetries of the standard model can be consistently combined. The upshot is that supersymmetry appears to be the only mechanism for unifying general relativity with the standard model (Haag (1975)).

There is no direct evidence that supersymmetry exists. None of the super-partners have been experimentally observed. Future experiments performed at the Large Hadron Collider (LHC) cannot definitively rule out all supersymmetric extensions of the standard model. But, for supersymmetry to solve the problems for which it was invented, at least a few of the super-particles should not be too heavy. To constitute dark matter, for example, the super-particles need to weigh no more than a few tenths of 1 TeV, which is well within the capabilities of the LHC. Another reason for the super-particles to be light is that a Higgs mass of around 0.125 TeV, the mass at which it was discovered, implies the super-particles should have relatively light masses. Unfortunately, if super-particles exist, they must be even heavier than many physicists had hoped. The situation is that several experiments have already been conducted at the LHC, and several "easy" models should have shown up experimentally but didn't. If the collider fails to find them, supersymmetry's usefulness could fade. That would be a blow, not just to the idea of supersymmetry, but also to the more ambitious unified theories that presume it (Castelvecchi (2012)).

### 3.3.3 Technicolor

Technicolor theories postulate a new super-strong interaction that gives a more natural explanation of electro-weak symmetry breaking (the mechanism through which elementary particles acquire mass) that overcomes the hierarchy problem. However, despite the precise agreement of the electroweak theory with experiments at energies accessible so far, the theory of electro-weak symmetry breaking remains unknown. The Higgs mechanism predicts the existence of the Higgs boson. But the Higgs boson is "unnatural" because quantum mechanical fluctuations produce corrections to its mass that lift it to such high values that it cannot play the role for which it was introduced without delicate fine-tuning of the standard model parameters (Hill (2003)).

Instead of introducing the Higgs boson, technicolor models generate masses for the $W$ and $Z$ bosons through the dynamics of new gauge interactions. This new strong interaction leads to a host of new composite, short-lived particles called "leptoquarks" at energies accessible to the LHC. The approach is natural because there is no elementary Higgs boson and no fine-tuning of the parameters. The discovery of a Higgs-like boson with a mass of approximately 125 $GeV/c2$ is not predicted by technicolor models. But, the main issue with technicolor theories is that no new gauge bosons and leptoquarks have been experimentally observed. If the collider does not find the new particles, it is unlikely that technicolor theories could survive (Hill (2003)).

### 3.3.4 Grand Unified Theories (GUTS)

The standard model is based on three independent interactions, symmetries, and coupling constants. By contrast, GUTS, at high energies, merge the electromagnetic, weak, and strong interactions into a single interaction. GUTS postulate that the standard model symmetry groups are subgroups of a larger symmetry group with one unifying coupling constant. Compared to the standard model, all realistic GUTS are complicated and introduce additional fields and interactions or even additional dimensions of space to explain the observed fermion masses and mixing angles. Due to these complications, and lack of experimental evidence, there is no generally accepted GUT.

Most GUTS predict the existence of new particles in the form of quarks and leptons. But the predicted particles cannot be observed because detecting their masses is beyond the reach of current and foreseen collider experiments. Instead, experiments designed to detect traces of grand unification are performed indirectly by observing proton decay, the electric dipole moments of elementary particles, or neutrino properties. Some GUTS predict the existence of magnetic monopoles (Ross (1984)).

A few GUTS explicitly break the baryon number symmetry, allowing protons to decay via the Higgs particle, magnetic monopoles, or new

*X* bosons. Proton decay is one of the few observable effects supporting the validity of the various proposed GUTS. However, to date, there have been no observations that would support any particular GUT. In addition, magnetic monopoles have not yet been observed in Nature.

### 3.3.5 Preons

Preon theories attempt to find theoretical answers to the large number of arbitrary constants within the standard model. For example, preon theories postulate that quarks and leptons are composite rather than fundamental particles. But, up to now, scattering experiments have observed no such structures (Goshaw (2012), § 1, p. 12). Furthermore, interest in preons was primarily predicated on the assumption that the Higgs boson would not be discovered. Since the discovery of the Higgs, it is doubtful that preons theories will garner much interest in the future.

### 3.3.6 String Theory

String theory postulates that matter does not consist of point-like particles, as in the standard model, but rather tiny loops of string. Each harmonic of the string represents a different particle. From this beginning, the laws of physics emerge. General relativity, electromagnetism, and Yang-Mills gauge theories appear surprisingly (Tong (2009) § 0).

The initial string theory was the "bosonic type." Its mathematical viability requires 25 extra spatial dimensions. It predicts the existence of a particle called a "tachyon." The mass of a tachyon is represented by an imaginary number, implying that it travels faster than light. Such a particle has never been observed, and its existence would conflict with known physics. Furthermore, the theory requires strongly interacting massless particles that cannot be brought to rest. Finally, the theory leaves out fermions (Greene (1999), pp. 180-181). Consequently, bosonic string theories have never been considered serious candidates to replace the standard model.

But in 1970, Pierre Ramond included fermions in a supersymmetric string theory. The new theory had no tachyons, and there were only nine instead of 25 extra spatial dimensions. At about the same time, Andrei Neveu and John Schwartz found a superstring theory where the superstrings interacted, allowing consistency with special relativity and the quantum theory. Two years later, Joël Scherk, Tamiaki Yoneya, and Schwartz found that some of the massless particles in the theory behaved like gravitons (particles that carry the gravitational force). This discovery led to a breakthrough and immediately suggested that string theory could unite all interactions (Jones (2010), p. 171).

In 1984, the first of two momentous events would take string theory from obscurity to one of the more intriguing ideas in physics. In collaboration with Michael Green, Schwartz provided strong evidence that string theory was finite, freeing it from the mathematical anomalies that had plagued previous attempts at unification. Within weeks, some of the most ardent critics of string theory began working on it. Theoretical physics had struck gold, the discovery of a quantum theory of gravity that unified forces with matter (Woit (2006), pp. 149-153).

But very quickly, researchers realized that string theory was not unique. Instead of one, there were five equally consistent 10-dimensional superstring theories. Accepting string theory meant assenting to the idea of extra spatial dimensions and the requirements of supersymmetry. Without them, the hope of creating a theory that solved the hierarchy problem and agreed with Nature vanished.

In any realistic unifying theory, the extra dimensions must be curled up into small difficult-to-detect loops. The many ways of curling up the extra dimensions amounted to a new theory. A vast array of particles arise when strings can move and vibrate in the complicated geometry of six extra spatial dimensions. Constants are associated with the masses of the particles and forces. String theory added many more, not less, freely varying constants. But even with the added flexibility, string theorists could not produce a string theory consistent with the standard model (Woit (2006), pp. 174-186).

Things looked bleak until 1995, when, drawing on the work of several string theorists (Ashoke Sen, Chris Hull, Paul Townsend, Michael Duff, and Schwarz), Edward Witten suggested that the five known string theories ($Type\ I, Type\ IIA, Type\ IIB, heterotic\ SO(32)$ and $heterotic\ E8 \times E8$) were different aspects of a single underlying theory Witten named "M-theory." When the five supersymmetric string theories were analyzed, Witten discovered that an object in one superstring theory was a dual description of a different object in another superstring theory. Because the extra dimensions can come in different shapes and sizes, string theory can come in many variants (i.e., different universes, where the different shapes and sizes of the extra dimensions have different physics). However, Witten showed that many of these geometries result in the same physics (Hull (1994)).

One theorist who took great interest in the Witten lecture was Joseph Polchinski. He showed that for the dualities in string theory to work correctly, the theory must include not only strings but also objects of higher dimension, which he deemed "branes" - short for membranes.

Strings are subject to boundary conditions. Like fastening each end of the strings on a guitar achieves the desired harmonics, the strings in string theory needed something to fasten themselves to. Polchinski realized that to correctly treat the dualities, the five different 10-dimensional superstring theories required a class of extended objects upon which open strings can end. He called these objects "D-branes."

D-branes are dynamic objects where the environment of strings stretching between them becomes very rich. Polchinski connected specific string modes to a non-abelian gauge theory, the same type of theory as the standard model. Furthermore, he showed that it might be possible to define observables in space as is done in quantum mechanics (Polchinski (1995)).

Polchinski's discovery was both a blessing and a curse. Branes are additional features added to the background geometry of the strings. Their addition dramatically increases the number of background geometries upon which the string can live. Moreover, adding branes made it possible to describe some particular types of black holes within string theory (Jones (2010), § 11).

Imagine stealing an amount of hot gas from the Universe and dropping it into a black hole. Since the gas cannot escape from the hole's gravitational pull, the entropy in the Universe would decrease, violating the second law of thermodynamics. Saving the second law requires the black hole to gain whatever entropy the in-falling gas originally had. Accordingly, Stephen Hawking surmised that a black hole should emit a characteristic spectrum of thermal radiation so that the Universe's entropy does not decrease. He employed thermodynamic arguments to derive the "Bekenstein entropy" for black holes. Counterintuitively, Bekenstein entropy is proportional to the black hole's surface area rather than its volume.

Black hole entropy is closely associated with the holographic principle. In 1993, Gerard 't Hooft (with help from Leonard Susskind) showed that the amount of "information" a space contains might be related to the area of a region's boundary, not its volume (Susskind (1995)). In QFT, everything can be viewed as information. In short, the holographic principle amounts to the following two postulates:

1. A gravitational theory describing a region of space is equivalent to a theory defined only on the surface area that encloses the region.
2. The boundary of a region of space contains at most one piece of information per square Planck length.

String theorists have constructed models in which a black hole is described as a very long (massive) string. This model gives rough agreement with the expected entropy of a Schwarzschild black hole, although an exact proof of this has so far eluded its proponents. It is relatively easy to count the number of degrees of freedom quantum strings possess if they do not interact with one another. But for the black hole problem, gravity is an interaction, and so, if the "string coupling" is turned off, in other words, if gravity is turned off, no black hole could arise. Calculating black hole entropy requires working in a regime where string interactions exist, which is a much more complicated problem.

String theorists can solve the simpler case of non-interacting strings requiring supersymmetry. Sometimes, the entropy calculation for zero-string coupling (no interaction) remains valid when the strings interact. The challenge for string theorists is to create a black hole without breaking supersymmetry. This has been done by building black holes out of D-branes. Calculating the entropies of these hypothetical holes gives results that agree with the expected Bekenstein entropy. Unfortunately, the cases studied so far all involve higher-dimensional spaces. Therefore, they do not directly apply to black holes observed in our Universe (Polchinski (1995)).

String theorists must find a way of interpreting the extra spatial dimensions in the context of real-world physics. In 1997, Juan Maldacena, inspired by the holographic principle, proposed the anti-de Sitter conformal field theory correspondence (AdS/CFT). He conjectured that the Universe is a 4-dimensional boundary of a 5-dimensional space containing the same information. In Maldacena's AdS/CFT correspondence, he proposed a new duality between a gauge theory defined on a 4-dimensional boundary (three spatial dimensions and one time dimension) and a 5-dimensional region (four spatial dimensions and one time dimension) (Jones (2010), § 11).

The AdS/CFT correspondence, called the "Maldacena duality," is the conjectured equivalence between a string theory and gravity defined on one space and a QFT without gravity defined on the conformal boundary of this space (Jones (2010), § 11).

Such were the successes of string theory over about thirty years. All the forces and particles of Nature could be explained as characteristic vibrations and actions of strings on branes. It included gravity and unified the bosons with the fermions. The laws of motion of the particles were united with the forces. First, string theory has only two fundamental quantities: The string tension and the string coupling constant. Second, it is governed by one simple principle: Strings move to minimize the surface area drawn out by the stings moving in space-time. Third, there was evidence that the five 10-dimensional supersymmetric string theories were manifestations of a single theory (M-theory). Fourth, string theory gave insights into black hole entropy. And finally, there was a way of relating string theory to gravity and the gauge symmetries of QFT, giving a plausible connection to our Universe (Smolin (2007), § 5).

But there are obstacles to overcome. M-theory has yet to be defined. It remains a conjecture and attracts criticism for lacking predictive power and being incomplete and untestable. Witten suggested that a general formulation of M-theory will require the development of a new mathematical language. With no underlying foundation, making meaningful progress on string theory has become increasingly difficult.

John Schwartz and Michael Green provided strong evidence that string theory was finite. But the "finite theory" assertion has since come under scrutiny. Most string theorists point to a proof by Stanley Mandelstam as evidence that string theory is finite. But there does not seem to be general agreement that Mandelstam's proof is complete. Primarily, string theorists make calculations using perturbation approximations. Without M-theory, all the perturbation expansions in string theory must be proven finite or replaced with exact calculations (Smolin (2007), § 8).

While introducing branes brought a welcome amount of flexibility to string theory, it added several layers of complexity. First, adding branes increases the number of background geometries upon which the string can live (Jones (2010), § 11). This increase dramatically expands the number of possible string theories.

The general theory of relativity is background independent, which means that space-time is dynamic, not fixed. Einstein was concerned about this aspect of his theory. However, he could not believe that the Universe was like a malleable piece of clay. So, he added to his equations a term, the "cosmological constant." It canceled the effect of the expansions and contractions, essentially making the Universe static. However, after adding the cosmological constant, Hubble, among others, argued that the Universe was expanding, eventually leading to the big bang theory of the Universe. Once physical evidence confirmed Hubble's hypothesis, Einstein felt that including the cosmological constant in his equations was a mistake.

But all experimental and observational evidence today indicates that Einstein's equations require a non-zero cosmological constant. This necessity is primarily due to the existence of dark matter and energy, which is thought chiefly responsible for the expansion of the Universe. It also suggests that the value of the cosmological constant is positive.

The Maldacena conjecture is described in anti-de Sitter spaces. All universes described on anti-de Sitter spaces have negative curvature, and hence, have negative cosmological constants. Our Universe, evidently, cannot be described on an anti-de Sitter space.

All string theories are background dependent, leading to the "vacuum problem." In classical physics, a vacuum has zero energy. In QTFs, the classical zero-energy space is replaced by a "vacuum state," defined as the lowest energy density state of a collection of quantum fields. This is because quantum fields exhibit fluctuations everywhere in space, even in regions otherwise "empty" (i.e., devoid of matter and radiation). These zero-point fluctuations give rise to an enormous vacuum energy density which is believed to contribute to the cosmological constant (Rugh (2000), pp. 663-705). In string theory, the vacuum state is given by specifying the background conditions necessary to make the vacuum state what it is. Such conditions require specifying the 10 or 11-dimensional space-time configuration and other technical details. Unfortunately, there are many consistent choices for a background, maybe an infinite number. And each of these backgrounds comes with a large set of parameters that specify the size and shape of the background space-time (Jones (2010), § 17), giving rise to an enormous number of possible string theories.

The plethora of possible string theories raises the issue of identifying which string theory describes our Universe. Further, it requires explaining what all the other string theories represent. The circumstance has culminated in an abundance of unusual ideas. Not the least controversial idea is the notion of an "omniverse," the conceptual ensemble of all possible universes with all possible laws of physics. The definition of a "universe" includes one set of physical laws and constants that govern it. The idea of an "omniverse" includes multiple sets of physical laws and constants, each expressing a wholly or partially separate universe.

Some string theorists regard an omniverse as an acceptable alternative to the idea of a "Universe." A proponent of this view is Leonard Susskind. He argues:

> *"... Physicists always wanted to believe that the answer was unique. Somehow there was something very special about the answer, but the myth of uniqueness is one that I think is a fool's errand ... If there were some fundamental equation which, when you solved it, said that the world is exactly the way we see it, then it would be the same everywhere. On the other hand, you could have a theory which permitted many environments, and a theory which permitted many different environments would be one in which you would expect that it would vary from place to place. What we've discovered in the last several years is that string theory has an incredible diversity – a tremendous number of solutions – and allows different kinds of environments. A lot of the practitioners of this kind of mathematical theory have been in a state of denial about it. They didn't want to recognize it. They want to believe the universe is elegant – and it's not so elegant. It's different over here, it's that over here. It's a Rube Goldberg machine over here. And this has created a sort of sense of denial about the facts of the theory. The theory is going to win, and physicists who are trying to deny what is going on are going to lose."*

Susskind (2003) published *"The anthropic landscape of string theory,"* in which he publicly gave up the idea that a unique string theory would be discovered. Instead, he introduces the concept of "the landscape" of string theories, a vast number of mathematically consistent possible universes.

Susskind's musings, both controversial and interesting, raise two important issues:

1. Of the vast landscape of string theories, which one describes our Universe or our part of the Universe?
2. If the Universe is part of an omniverse, how is this verified?

Susskind's answer to the first question is the "anthropic principle." Many string theorists now champion using the anthropic principle to determine which string theory best describes the Universe/Omniverse.

What is the anthropic principle? It seems there are finely tuned connections between the values of the physical constants of a theory and the conditions necessary for life to exist. These connections are often referred to as "anthropic coincidences." Here are some of the most significant:

1. The electromagnetic force is 39 orders of magnitude stronger than the gravitational force, and if those forces were more comparable in strength, stars would have collapsed long before evolving.

2. The vacuum energy density of the Universe is at least 120 orders of magnitude lower than some theoretical estimates. If, at any time, it were as large as the calculations suggest, the Universe would have quickly blown apart.

3. The electron mass is less than the difference between the neutron and proton masses. A free neutron can decay into a proton, electron, and anti-neutrino, and if this was not the case, the neutron would be stable, and most of the protons and electrons in the early universe would have combined to form neutrons, leaving little hydrogen as the main component and fuel of stars.
4. The neutron is heavier than the proton. Still, neutrons are not so much heavier that they cannot be bound in nuclei, where energy conservation prevents the neutrons from decaying. Without neutrons, there would not be the heavier elements needed for building complex systems such as life.

5. The carbon nucleus has an excited energy level of around 7.65 $MeV$, and without this, insufficient carbon would be manufactured in stars to form a basis for life.

The anthropic principle comes in two versions:

1. **The weak anthropic principle (WAP)**: Our location in the Universe is necessarily privileged to the extent of being compatible with our existence as observers.

2. **The strong anthropic principle (SAP)**: The Universe (and hence the fundamental parameters on which it depends) must admit the creation of observers within it at some stage.

Even the string theorists who now embrace the anthropic principle, such as Susskind and Polchinski, once despised it as totally unscientific. The unscientific criticism arises because the SAP is sometimes invoked as an argument for a supernatural designer of the Universe. Ironically, the WAP is often used as an argument against a supernatural designer, as Susskind does in his book (Jones (2010), § 11).

The most divisive issue associated with the anthropic principle is the idea of a finely-tuned universe. A fine-tuned universe is one where the conditions that allow life to exist occur only when certain fundamental physical constants lie within a narrow range of values. If any fundamental constants were only slightly different, the Universe would not be conducive to establishing and developing matter, astronomical structures, elemental diversity, or life (Isaak (2005)).

The existence and extent of fine-tuning in the Universe are disputed. Fine-tuning does not require multiple universes. What it does require is an explanation of why the critical constants of Nature have the values they do.

Another possibility is that the parameters are not fine-tuned but random. The concept of multiple universes can explain why our Universe seems fine-tuned for conscious life. If there were a large number (possibly an infinity) of different physical laws (or fundamental constants), some of these universes would have laws suitable for stars, planets, and life to exist. The makeup of our universe is then a consequence of the laws of probability (Jones (2010), § 11). This position is essentially the one that Susskind adopts.

But those who make the "no fine-tuning" argument must address the vacuum energy problem. This issue is essentially the cosmological constant problem since any vacuum energy density is equivalent to this parameter in Einstein's theory of general relativity. Calculations give a value for the vacuum energy density of some 120 orders of magnitude greater than the maximum value possible from observations. Since this density is constant, it would seem that its value is fine-tuned for the existence of life (Isaak (2005)).

Until recently, the cosmological constant's value was considered to be zero. However, in 1998, two independent research groups studying distant supernovae discovered that the expansion of the Universe was accelerating. More recent observations have confirmed this result. The source of this cosmic acceleration is believed to be some still unidentified dark energy, which constitutes 70 percent of the mass of the Universe. One possible explanation is gravitational repulsion employing the cosmological constant through the vacuum energy field allowed by general relativity. If this is the case, the cosmological constant seems fine-tuned (Weinberg (1987), pp. 2607–2610).

# Chapter 4: The Practical Challenges of Unification

***"What I am going to tell you about is what we teach our physics students in the third or fourth year of graduate school... It is my task to convince you not to turn away because you don't understand it. You see my physics students don't understand it... That is because I don't understand it. Nobody does."***

***— Richard P. Feynman, QED: The Strange Theory of Light and Matter***

## 4.0 Introduction

In Chapter One, we discussed how two incompatible theories dominate physics. "Quantum gravity" is the name for a future theory that, if discovered, would unite quantum theory with relativity. However, discovering such a theory faces both practical and conceptual obstacles, which, up to now, have thwarted attempts at finding a generally accepted solution.

In Chapters Two and Three, we discussed the discoveries that ultimately led to the standard model of particle physics and the general theory of relativity. While arguably the most successful endeavor in the history of modern science, as a unifying theory, the standard model harbors shortcomings, most notably, a theory of gravity. Its theoretical basis has not been proven consistent, and it requires several experimentally derived inputs that remain theoretically unexplained. Outside the domain of the energy scales where it's been tested, its predictions are clearly wrong. The input parameters in the model require unnatural fine-tuning, and artificial mechanisms are needed to make the model's predictions agree with observations.

General relativity is our current theory of gravity. Given its agreement with experimental results and ability to predict new phenomena, it is perhaps the most successful purely theoretical development in the history of science. The theory explains most of the physics at large scales. Moreover, it has continued to explain new observations that it was not designed to predict, including foretelling the existence of black holes, and is consistent with recent observations in cosmology. But it applies only to large-scale physics and cannot be extended to small-scale physics.

For these and other reasons, physicists look for an alternative theory that can unite the standard model with the theory of relativity - string theory being the most promising option. However, while ingenious, the alternatives come with their own set of problems. In this Chapter, a discussion of the practical challenges in overcoming the apparent dichotomy between relativity and quantum mechanics is undertaken.

## 4.1 Quantum vs. Classical Theories

The theory of relativity is a classical theory. Three features of quantum theories distinguish them from classical theories, and one feature illustrates how the two are connected (Bohm (1951), § 2):

1. The indivisibility of quantum processes.
2. Incomplete determinism in quantum laws.
3. The holistic nature of quantum processes.
4. The correspondence principle.

### 4.1.1 The Indivisibility of Quantum Processes

Classical theories assume that all physical processes happen gradually. There are no sudden jumps or spikes in energy transitions, for example. Mathematically, gradual means smooth transitions from one state to the next. But experiments have shown that such an assumption is invalid (Bohm (1951), § 2). For example, Bohr's theory of the atom, the photoelectric effect, and the absorption and emission of radiation in black bodies illustrate that quantum processes come in discrete packets (i.e., there is a limit to how small these packets can be).

The indivisibility of quantum processes influences the mathematical approach to describing quantum phenomena. Classical theories are based on "infinite divisibility," which assumes that a process, like the trajectory of a rocket, can be divided into segments of any desired smallness. For example, the differential and integral calculus describing an object's motion requires smooth transitions from one state to the next. But applying the calculus to discretely discontinuous processes is challenging, where state changes happen uncontrollably, and predicting the disposition of the following state is impossible.

### 4.1.2 Incomplete Determinism in Quantum Laws

Classical physics is causal. If the system's current state is known sufficiently, all future states can be predicted with certainty. The current state of a system determines, in a sense, all future states. All that is needed is the following:

1. The system position at any instant in time.
2. The velocity at that time.
3. The value of the forces at all times.

This characteristic of classical physics is called "determinism." But numerous experiments have shown that quantum processes are uncontrollable. For example, it is impossible to predict when energy transfers will occur (Bohm (1951), § 2). Only betting odds can be given. The quantum world is probabilistic, not deterministic.

The probabilistic nature of quantum phenomena is one of the most controversial aspects of the theory. Classically, a problem requires a statistical treatment if the many variables involved cannot be easily accounted for. But, if those variables could be known, deterministic solutions follow. But "probability," in the quantum sense, plays the role of a Natural law. There are no "difficult to account for" variables that, if known, would allow a deterministic solution. Therefore, processes at the quantum level are not more fundamentally discoverable (Bohm (1951), § 2) and are not knowable beyond the limits of probability.

Many physicists at the time, notably Einstein, took issue (some still do) with this feature of quantum mechanics. In the now-famous EPR paper, Einstein and his collaborators argued that the quantum theory ignored variables that could explain the nature of atomic phenomena without some aspect of the system becoming uncontrollable. He felt this would return determinism to the atomic realm. His objections led to Bell's theorem (more on this later). And Bell's theorem, as it turned out, was accessible to experimental verification. Originally designed to validate the theorem, experiments showed that Bell's theorem was violated, as predicted by the quantum theory. Although the experiments are not entirely conclusive, the results imply that it is not a lack of information that prevents a deterministic explanation of the nature of quantum processes. This result gave credence to the notion that the quantum theory is consistent with its current interpretations.

The probabilistic nature of the quantum theory determines how quantum processes are described mathematically to some extent. For example, it is impossible to specify a particle's exact location and momentum simultaneously. Such variables, called "conjugate variables," are a consequence of the uncertainty principle, which forbids simultaneously knowing the exact values for these variables. In quantum problems, conjugate variables are considered uncontrollable. At best, only statistical averages can be calculated for them. In experiments where conjugate variables are essential, a slightly different value is obtained each time the experiment is conducted. But a sufficient number of experiments will lead to values hovering around a statistical average. Classical equations containing conjugate variables cannot carry over into the atomic realm. Any problem requiring complete knowledge of the status of a system at every moment will find no place at the quantum level (Bohm (1951), § 2).

### 4.1.3 The Holistic Nature of Quantum Processes

The "*particle*" and the "*wave*" represent the two great concepts of classical physics. Virtually every problem in the classical realm can be formulated in terms of one or the other of these concepts (Halliday (1988), § 17.1). The two concepts, however, are quite different. Particles have a center of mass located at a particular space-time location. Waves, on the other hand, are spread out with no specific location in space and time. Moreover, waves display interference, diffraction, polarity, and periodicity - attributes that particles were thought not to possess. Classical physics classifies phenomena into either "wave-like" or "particle-like" (Bohm (1951), § 3), but phenomena are never both wave-like and particle-like simultaneously.

But, de Broglie's equation, $p = h/\lambda$, which relates momentum (a feature of a particle) to its wavelength (a feature of a wave), suggests that treating waves and particles as distinct phenomena is a mistake. But then, how should matter be described if matter is both particle-like and wave-like?

### 4.1.4 Wave-Particle Duality

Eventually, the explanation physicists settled upon was "wave-particle duality." The idea arose primarily through the study of light. What, it was asked, is light? Newton (1998) thought it consisted of tiny particles interacting with the eye, producing images. However, experiments conducted by Thomas Young showed, conclusively, that light was a wave. Likewise, Maxwell's theory of electromagnetism describes light as an electromagnetic wave. Maxwell's wave description of light went unchallenged for just short of a century. But late in the nineteenth and early twentieth century, experiments showed that light displayed characteristics of a particle under certain conditions (photoelectric effect).

Suppose two small slots are cut horizontally into an armored plate in close proximity. If a machine gun sprays a barrage of bullets at the plate, in what manner will the bullets pass through the slots? This problem is relatively complicated, as some bullets will ricochet off the edges of the slots in unpredictable directions. Furthermore, the number of bullets passing through the slots will depend on the size and shape of the bullets and the slots. Closing off one of the slots and setting up a detector behind the plate will reveal where each bullet strikes. Not surprisingly, the most probable location is directly across from the open slot. At positions farther away from this central location, bullets will strike less frequently. From a probability standpoint, the pattern that emerges on the detector is called a "probability distribution." If the experiment is repeated many times, similar results occur. This information can be used to predict the probability that a bullet (think particle) strikes the detector at a specific location.

What happens if both slots are open? The pattern does not change. The impact pattern shows a similar probability distribution. The most likely place a bullet will strike the detector is directly across from each opening. Bullets strike off-center positions less often (Figure 4.1.4-1).

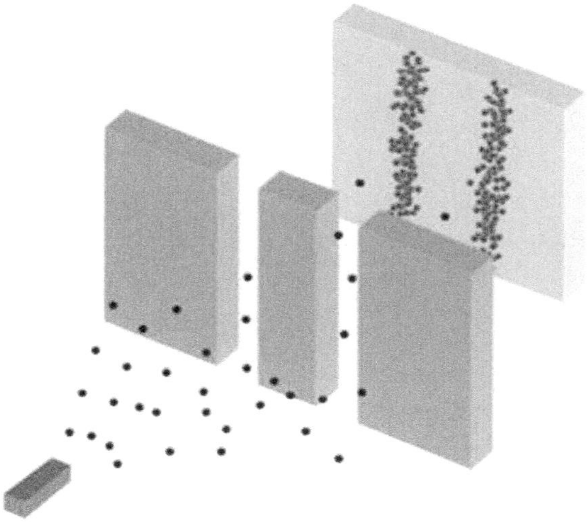

The two-slot bullet experiment.
Figure 4.1.4-1.

Imagine conducting a similar experiment using a light source. What happens when a light beam passes through two small slots of an adequately designed apparatus? If light is wave-like, it should show different characteristics than a particle. There would be no center of mass. Instead, waves spread out over large areas. Characterized by two opposite poles, crests and troughs, waves fluctuate up and down or back and forth as they move through space and time. When two waves meet, the vacillating motion can cancel or be reinforced depending on whether a crest of one wave meets the trough of another to cancel or two crests or two troughs meet, adding together to create a larger crest or trough.

If one of the slots is closed, the pattern on the detector resembles the bullet experiment - a distinct peak. But when both slots are open, the light waves pass through both slots at once; in phase, they reinforce; out of phase, they cancel, creating an interference pattern on the detector (see Figure 4.1.4-2).

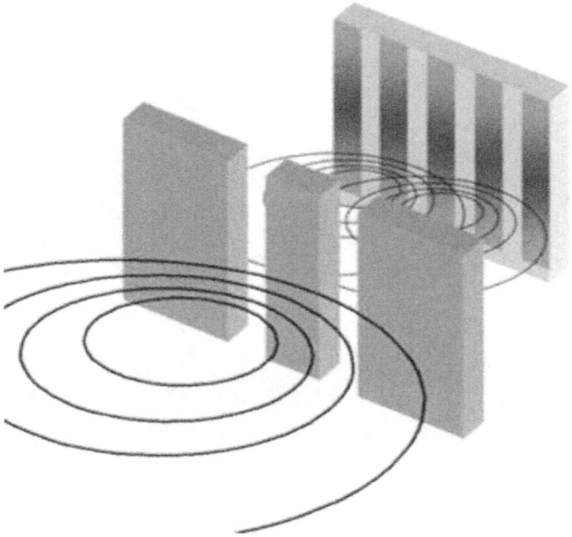

The two-slot light wave experiment.
Figure 4.1.4-2.

Lighter bands appear where the waves have reinforced, while darker bands appear where the waves cancel. The bands show more intensity in the center part of the detector but less intensity at locations away from the center. If one slot is closed and the other remains open, the interference pattern vanishes, and a pattern emerges similar to the bullet pattern. This pattern occurs because only one wave is produced, so there is no reinforcement or canceling effect created by interfering waves.

The early twentieth century saw the invention of the "electron gun," essentially a vacuum tube, the same kind used in the first black and white television sets. At the time, there was little doubt that an electron was a particle. It had mass, a particular position in space, momentum, and energy, just like any other particle. Moreover, electrons can transfer their momentum and energy just like other particles.

Experiments that amounted to the two-slot experiment, except with an electron gun, were conducted. With one slot closed, the results were virtually identical to the bullet experiment. Electrons with enough momentum dislodged other electrons from the detector. But when both slots were open, electrons not only dislodged other electrons, but an interference pattern appeared - areas where no electrons were dislodged. This result was different from the bullet experiment. In classical physics, particles did not act like waves and vice versa (see Fig 4.1.4-3).

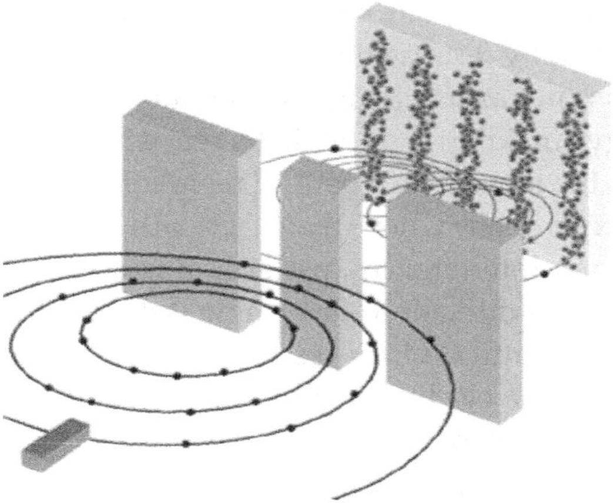

The electron gun experiment.
Figure 4.1.4-3.

Surprisingly, an explanation for the wave-particle paradox that would satisfy all or most of the influential physicists of the day did not materialize. Instead, after much discussion and theorizing, the explanation invoked a debate.

The electron gun experiment was modified and repeated to gain a deeper understanding of the nature of matter. The experiment was slowed down so that only one electron at a time was emitted from the gun, ensuring that the electron would pass through only one of the two slots. But the modification did not change the results. When both slots were open, an interference pattern emerged.

Again, the experiment was modified by directing a light beam just behind the openings to observe which slot the electron passed through. A flash of light appeared every time an electron passed through one of the slots. In this case, electrons behaved like particles - 50% of the time passing through one slot and 50% of the time through the other. Electrons were never seen passing through both slots simultaneously. But something unexpected happened. The interference pattern disappeared.

Evidently, the directed light beam caused a disturbance, which affects qualitatively and irreversibly the behavior of the electrons. It is possible to minimize this disturbance by increasing the wavelength of the light beam, thereby decreasing its energy so that the beam's impact on the electrons is not as severe. But just about the time the light lacks sufficient energy, making it difficult to see which slot the electron passes through, the interference pattern reemerges. As yet, no one has conducted an experiment where the exact path of an electron can be known, and an interference pattern appears simultaneously. Consequently, it was difficult to determine if the electron acted like a wave or a particle until and unless it interacted with a measuring device. Seemingly, the electron itself has no definition, but its nature is determined by its interaction with a measuring device. This is the conundrum of wave-particle duality.

### 4.1.5 Schrödinger's Cat

The confounding experimental results at the quantum level led physicists to face the grim reality that matter was not made from the simple substances described by classical theories. The only choice was to find as reasonable an interpretation as possible that would suffice to describe the constitution of matter.

The disposition of a quantum mechanical system is called a "configuration of states" or a "configuration space." Schrödinger equation is a deterministic equation because, given the configuration space, the time evolution of that space is determined precisely by Schrödinger's equation. Mathematically, Schrödinger's equation follows the "law of superposition," which says that if $\psi_1$ and $\psi_2$ are both solutions to Schrödinger's equation, then so is $\psi_1 + \psi_2$ (Schrödinger (1926)).[5]

To better understand the idea of a quantum mechanical state, consider the parable of "Schrödinger's cat." A cat is placed in an opaque box, so it is impossible to see inside without opening it. The box is equipped with a tube split into two branches; one branch leads to a capsule filled with sleeping gas, and the other leads away from the gas. An electron is released into the tube. If the electron follows the branch leading to the gas capsule, the interaction will cause the capsule to burst, releasing the gas and rendering the cat unconscious. The cat remains conscious if the electron follows the path away from the gas. The cat's configuration space consists of two states: $\psi_1 \equiv$ "*conscious cat*", $\psi_2 \equiv$ "*unconscious cat.*"

After the electron is released into the box, is the cat in the state, "$\psi_1$" or "$\psi_2$?" There is only one way to find out; open the box and look inside. The cat will be found either unconscious or conscious.

---

[5] Strictly speaking, the most general solution is $c_1\psi_1 + c_2\psi_2$, where $c_1$ and $c_2$ are constants, but these technical details are not important for the discussion here.

But Schrödinger equation follows the law of superposition. If $\psi_1$ and $\psi_2$ are solutions to the equation, so is $\psi_1 + \psi_2$. In terms of the cat problem, what does "$\psi_1 + \psi_2$" represent? Evidently, it indicates that the cat is in some alternating state of unconsciousness and consciousness. However, quantum mechanical systems resist any attempt at observing them in superposition (Penrose (2005), § 29.7). When the box is opened, the cat will be either unconscious or conscious, not in some nebulas state between unconsciousness and consciousness. Opening the box represents the "collapse of the wave function." An observation will, in some sense, cause a quantum system to jump to one of the possible single states. In the cat problem, this involves jumping to either $\psi_1$ or $\psi_2$.

The superposition law represents the lack of complete determinism in the quantum theory. In Schrödinger's cat parable, it cannot be determined whether the cat is unconscious or conscious without opening the box. Only betting odds can be given – 50% chance of being unconscious, 50% chance of being conscious. The cat's state is unknown until the box is opened and an observation is made. Likewise, the quantum theory resists any attempt at determining, with certainty, if a system will show wave-like or particle-like properties until it interacts with a measuring device.

Classical physics separates Nature into distinctly definable parts, the observer and the observed. Problems are formulated by distinguishing between observer and observed and assuming that the observer's impact on an experiment has little significance and can be ignored. The quantum theory resists any such distinction. Systems are thought about holistically. There is no way of breaking problems into distinctly definable parts. It is impossible to ignore the effect of observation on any given experiment (Bohm (1951), § 2).

### 4.1.6 The Correspondence Principle

At this point, quantum mechanics exhibits three non-classical characteristics: (i) State changes, such as energy transfers, come in discrete units ($E = h\nu$) rather than continuous state changes, (ii) only the probability of an energy transfer can be determined between states, and (iii) it is not possible to break a problem into distinct parts (Bohm (1951), § 2).

But, if quantum mechanics successfully describes microscopic objects, such as atoms and other elementary particles, and classical physics explains macro-systems, like springs and capacitors, how are these incompatible approaches reconciled?

Quantum mechanics answers this question through the "correspondence principle." First formulated by Bohr, it states that the laws of quantum physics must, if many quanta are involved, agree with classical results on average. The "many quanta are involved" statement is usually called "in the classical limit."

Bohr's correspondence principle is very general. The key phrase is "in the classical limit." When Bohr adopted the principle, it was unclear precisely what he meant (Bokulich (2010), § 4). It is still the subject of research that goes beyond the scope of this book.

From a unification standpoint, the correspondence principle is unsatisfactory. The fact that, when many quanta are present, classical methods can be employed for obtaining satisfactory solutions to physical questions implies that the errors encountered are sufficiently small and can be ignored. The argument does nothing to address the more profound philosophical divergence between how classical and quantum theories describe the nature of matter.

## 4.2 The Problems of Measurement and Interpretation

Two diametric processes characterize quantum mechanics. First, there is the unitary process described by Schrödinger's equation, whereby a superposition of quantum states evolves in time in a determined manner. The second process is a quantum state reduction that produces a specific outcome, which occurs as a result of a measurement. Roger Penrose called these the "U" and the "R" processes. How an outcome, R, can develop from the process, U, is the crux of the measurement problem (Penrose (2005), § 22).

### 4.2.1 The Copenhagen Interpretation

The Copenhagen interpretation is one of many ways to explain the mathematical formalism describing atomic experimental results. While Heisenberg was employed as a lecturer in Bohr's Institute for Theoretical Physics at the University of Copenhagen, during their collaborations, they, along with Max Born and others, formulated the "Copenhagen interpretation" of quantum mechanics.

There is no official "Copenhagen interpretation." It will not be found in any formal document. This is because Bohr and Heisenberg disagreed on some critical aspects of how the experimental results and equations that describe them should be interpreted (Bokulich (2010), § 4.3). But roughly speaking, the Copenhagen interpretation adheres to the following principles:

1. **The Completeness Principle**: A system is completely described by a wave function, $\psi$, representing an observer's knowledge of the system.

2. **The Indeterministic Principle**: Descriptions of Nature are probabilistic, and the probability of an outcome is related to $\psi^2$ - the square of the wave function.

3. **The Conjugate Variable Principle**: It is impossible to know the values of all the system's properties simultaneously; those not known precisely can only be described probabilistically.

4. **The Holistic Principle**: Matter can only be described as wave-like or particle-like after interacting with a measuring device (e.g., after an observation has occurred); trying to describe matter as wave-like or particle-like before a measurement is not possible.

5. **The Correspondence Principle**: Quantum mechanical predictions involving large systems (many quanta) should, on average, agree with classical predictions.

How does the Copenhagen Interpretation correspond to the electron gun experiment discussed previously (Sec. 4.1.4)? First, the electron must pass through one of the openings. Hence, the probability it passes through one of the two openings is 50%. Conversely, there is a 50% chance it passes through the other opening. This is a reflection of the completeness principle (Principle 1). The electron must pass through one of the two openings (50% + 50% = 100%). It cannot be determined *apriori* which of the two openings the electron passes through, reflecting the system's probabilistic nature (Principle 2).

Principle 4 is the most controversial. It involves the collapse of the wave function. If no directed light beam is projected upon the slots, then the probability waves recombine at the detector to form the interference pattern. In other words, the electron behaves like a wave. But if a light beam is shined on the slots to see which opening the electron passes through, the wave function ceases to be a factor, and it will be known with certainty which opening the electron passes through. There is no chance it passed through the other opening. In this case, the electron acts like a particle, and the interference pattern on the detector vanishes.

Max Born first suggested that the waves associated with the wave function, $\psi$, should not be interpreted as waves propagating through a material substance, like water or sound waves. Instead, they should be interpreted as measures of probability. Waves of probability are related to the uncertainty principle in the sense that, what individual waves will do, cannot be predicted with certainty. Instead, only betting odds can be given.

Such an interpretation represented a complete break from the determinism of classical physics. It meant that the outcome of any experiment was only statistically predictable. The exact nature of each electron can never be determined. Those experiments designed to measure wave behavior will see the electron as a wave. Those experiments designed to measure the particle properties of the electron will see the electron as a particle. No experiment can simultaneously measure the wave and particle properties of the electron. Known as the "complementary" aspect of quantum mechanics (Schombert (2012)), this interpretation is the fundamental notion behind "wave-particle duality."

Adopting the Copenhagen interpretation of quantum phenomena created a sharp divide between the determinism of classical physics and the new quantum physics. The divide spilled over into debates between those who accepted the new interpretation and those who didn't. Counted amongst those who didn't include many eminent physicists of the day, including Einstein, Schrödinger, and de Broglie. They were pitted against Bohr, Heisenberg, Born, and others, who more or less adopted the new physics. Einstein, in particular, could not accept the indeterminacy of the quantum theory. In his famous debates with Bohr, he was said to have complained that "God does not play dice ... " to which Bohr reportedly replied, "Stop telling God what to do ...."

### 4.2.2 The Measurement Problem

Under the Copenhagen interpretation, no physical reality is associated with the U process. Instead, it simply gives the probability that a potential outcome materializes.

The measurement problem in quantum mechanics is the unresolved issue of how (or if) wave function collapse occurs. In other words, what is the nature of the transition from U status to R status? The inability to observe this transition directly leads to different interpretations of quantum mechanics. The wave function in quantum mechanics evolves according to Schrödinger's equation into a linear superposition of different states (U process). Still, actual measurements always find the system in a definite state (R process). Therefore, any future evolution is based on the system's state after a measurement, implying that the measurement "does something" to the process under examination (Penrose (2005), § 22.1). Whatever that "something" might be is not explained by quantum mechanics.

The measurement problem is not unlike the torturous grappling by 17th-century philosophers who tried vainly to explain how a "non-material" idea could cause a physical reaction - like when a student raises a hand in response to a teacher's question.

Returning to the parable of Schrödinger's cat, under the U process, the cat is evolving into a linear superposition of states characterized by "*unconscious cat*" and "*conscious cat*" (the cat seems to be in a "mixed" state) unconscious and conscious simultaneously. However, a particular observation of the cat finds it either unconscious or conscious. How are the probabilities associated with the U process converted into a sharply well-defined reality associated with the R process (Penrose (2005), § 29)?

According to Bohr, this jump represents a simple increase in the observer's knowledge. When an observation is made, the observer becomes aware of the actual situation and sees the cat as unconscious or conscious. Many physicists find this explanation unsatisfying. According to Bohr's interpretation, the transition from the U to the R status occurs wholly in the observer's mind. The cat does not physically jump from its U state to the reality of being unconscious or conscious. It is unclear whether quantum mechanics has anything to do with the mind of an observer. The U process, while not physical, gives the probabilities of tangible outcomes, and the R process represents a tangible outcome. What did either of these processes have to do with the mind?

The analysis suggested that Bohr's version of quantum mechanics was deeply flawed. While it does describe what happens when an observer takes a measurement, the observer and the act of measurement are treated classically, which is undoubtedly wrong. The observer and the measuring device are made of the same materials (protons, neutrons, and electrons) that all other things are made of. Therefore, they should be governed by a wave function, which is entirely missing from Bohr's explanation. Physicists and their apparatus must be governed by the same quantum mechanical rules that govern everything else in the Universe. These rules are expressed in terms of a wave function (or, more precisely, a state vector) that evolves perfectly deterministically. So, where do the probabilistic rules of the Copenhagen interpretation come from? The shortcomings in Bohr's interpretation led to efforts to find a better explanation (Weinberg (2005), p. 31).

In opposition to Bohr, Wolfgang Pauli and Heisenberg argued that the observer produced the collapse. Pauli described quantum mechanics as lucid mysticism (Marin (2009), pp. 807–822). The involvement of consciousness in the collapse of the wave function can be summarized:

> *The rules of quantum mechanics are correct but there is only one system (one wave function) which may be treated with quantum mechanics, namely the entire material world. There exist external observers which cannot be treated within quantum mechanics, namely human minds, which perform measurements on the brain causing wave function collapse* (Schreiber (1995)).

This interpretation attributes wave function collapse to consciousness itself. Specifically, a non-physical mind is postulated as the only measurement apparatus (Schreiber (1995)). Since consciousness is not subject to physical law, it is not quantum mechanical. The "consciousness causes collapse" argument appears to overcome the objection that observers and measuring devices should be treated quantum mechanically.

This viewpoint leads to deep philosophical questions. In many philosophies, the conscious mind is conceived as a separate entity, existing in a realm not described by physical law. This perspective was first introduced in detail by Descartes in the 17th century. Those who support this view maintain that the physical world described by quantum mechanics suggests a kind of mind/body dualism. Of course, it would seem that the proponents of this way of thinking would face the same difficulties in defending it as did Descartes. He failed to adequately describe the mechanism which allowed a non-physical mind to affect the actions of a physical body, which it indeed does. This interpretation of the quantum theory adds nothing to resolve this age-old question.

Beyond that, as Hugh Everett points out (Everett (1957)), difficulties arise when the Universe is conceived as possessing more than one observer. Depending on their perspectives, observers do not necessarily perceive identical realities even when witnessing the same physical system. If consciousness causes us to perceive reality, each observer, owning their consciousness, should perceive a reality that is, at least in part, subjective. But then, why do some observers not see a green moon, others a red moon, and still others a yellow moon, for example?

Evidently, the answer lies in the meshing of quantum mechanics with Eastern Mysticism, including Hinduism, Taoism, and Buddhism. These beliefs include a transitory interconnectedness in all of Nature - a shared consciousness and the experience of one reality. According to this belief, there is a single consciousness shared by all. And this, it seems, overcomes the problem of having numerous realities embodied in many conscious observers.

However, this point of view was difficult to square with the other aspects of physics. While the "consciousness causes collapse" doctrine holds an essential role for the conscious mind, it is difficult to explain how the earlier expansion of the universe evolved those conscious minds. As John Stewart Bell said (quite sarcastically), "Was the wave function waiting to jump for thousands of millions of years until a single-celled living creature appeared? Or did it have to wait a little longer for some highly qualified measurer - with a Ph.D.?" (Bell (1981), p. 611).

### 4.2.3 The Many Worlds Interpretation

One way to handle the measurement problem is to deny wave function collapse. Hence, the problem of explaining the mechanism by which wave function collapse occurs vanishes. And this is what Hugh Everett (1957) proposed. He did this by merging the micro and macro worlds. Observers became an integral part of the system by introducing a universal wave function that included observers and the objects observed as parts of a single quantum system. In Everett's system, the continuous evolution of a wave function is not interrupted by acts of measurement. Everett's conjecture became known as the "many-worlds" interpretation. It asserts that the universal wave function is objectively real and denies wave function collapse, implying that all the possibilities embodied in the wave function materialize as alternative worlds or universes. The universal wave function contains branches for every possibility. Each branch has a copy of the observer, and each copy perceives one of the possibilities as the outcome. According to a fundamental property of Schrödinger's equation, once formed, the branches do not influence one another. Each branch embarks on a different future independently of the others. This branching or splitting of the quantum possibilities into independent worlds, each having its own classical reality, has generally become known as "decoherence theory." The many-worlds interpretation views reality as a multi-branched tree wherein every possible quantum outcome is realized (Byrne (2007)).

There are two principal objections to the many-worlds interpretation. First, it is unclear what mechanism causes the splitting or acts of decoherence to happen. The present-day understanding of decoherence is not a precise, self-contained explanation. There are many interpretations of the quantum theory based on different versions of decoherence. And it is unclear which version is the correct one. Secondly, each of the many worlds is self-contained. There is no interaction between them, which raises questions about how the many-world conjecture can make testable predictions, subject to falsification. In that sense, the many-world interpretation does not buy much in the way of solving the measurement problem. On the one hand, it removes the necessity of explaining wave function collapse. Still, it gives back that advantage by postulating a multiverse of worlds, only one of which is there any direct knowledge.

### 4.2.4 Environmental Decoherence

The many-worlds interpretation shares similarities with later "post-Everett" interpretations of quantum mechanics that use decoherence to explain the measurement process. However, while the "many-worlds" hypothesis treats the alternative worlds as authentic, other decoherent interpretations regard the extra quantum worlds as metaphorical or are agnostic about their reality (Schlosshauer (2005)).

An interpretation that has gained popularity recently is called "environmental decoherence." Environmental decoherence attempts to explain the transition from the quantum U process to the classical R process by analyzing the interaction of a system with a measuring device or with the environment. To clarify, imagine a quantum mechanical particle or system of particles as an isolated system floating in empty space. This simplification may be acceptable in some cases, but there is no such thing as an isolated system in the real world. Typically, a particle in flight will collide with air molecules or will emit thermal radiation that gets absorbed by the environment. Any interaction with the environment leads to an entanglement between the particle's state and the environment. As the entanglement diffuses throughout the environment, the total state can no longer be separated into the direct product of a particle "state" and an environment "state." What was once a superposition of particle states becomes a superposition of particle and environmental states. At this point, the particle ceases to be in a quantum superposition of states. Instead, it is in a statistical ensemble of states.

The decoherence process results in the particle appearing to have collapsed in a manner described by the Born probability law. Born's law then becomes a consequence of the interaction between the particle and the environment. In essence, unlike the many-worlds interpretation, environmental decoherence involves at least the appearance of wave function collapse caused by the particle interacting with its environment.

Recall that a quantum mechanical superposition of states fundamentally differs from a classical ensemble of states. In classical physics, an object is never both "here" and over "there." Objects are never in two places at once. So, in classical physics, the probability of an object being both "here" and over "there" is zero. However, recall the two-slot interference experiment where electrons pass individually (one at a time) through a double slot. Within the standard quantum mechanical formalism, the electron cannot be explained by a single wave function but requires two wave functions to describe the passage through a particular slot ($\psi_1$ or $\psi_2$), but, then, only by the superposition of these wave functions (i.e., $\psi_1 + \psi_2$). The correct density distribution of the pattern on the screen is not given by the sum of the squared wave functions individually (i.e., $P \neq |\psi_1|^2 + |\psi_2|^2$), but by the square of the sum of the individual wave functions ($P = |\psi_1 + \psi_2|^2$). So, from a quantum mechanical standpoint, the electron is treated as if it were in two places simultaneously. Of course, the "two places at once" superposition state of the electron is not observable. The Copenhagen interpretation maintains that upon measurement, the electron "jumps" from a superposition of states to one of the singular states, $\psi_1$ or $\psi_2$.

In opposition to the Copenhagen interpretation, environment decoherence argues that when a system is in a superposition of states ($\psi_1 + \psi_2$), it becomes entangled with its environment and quickly decoheres into an ensemble of classical states. In other words, it quickly devolves from $P = |\psi_1 + \psi_2|^2$ to $P = |\psi_1|^2 + |\psi_2|^2$. A well-defined mathematical process called a "density matrix" describes this transformation. The classical states can then be given in terms of classical probabilities, where the probability of observing the electron in superposition quickly becomes zero.

Environmental decoherence has become popular with many physicists because it effectively removes the mysticism of the "observer causes collapse" interpretation. And because the system and the environment are both treated quantum mechanically, it addresses the objections to Bohr's interpretation. Because of this, it is often argued that environmental decoherence effectively solves the measurement problem.

There are about as many opinions on this question as different approaches to environmental decoherence. Therefore, describing environmental decoherence as a program rather than a single technique is more accurate. The critical question seems to be what constitutes the "environment" and what constitutes the "system." The many variants of environmental decoherence treat these two key components differently.

Whether or not environmental decoherence solves the measurement problem largely depends on a particular point of view. Physicists not overly concerned with ontological interpretations of physical events will likely be more inclined to answer the question in the affirmative. The issue is the nature of the density matrix, which is the mathematical tool for proceeding through the environmental decoherence process. Density matrices are not unique. The same density matrix can describe more than one system. For example, while a system composed of "spin-up" and "spin-down" electrons in an EPR environment is certainly ontologically different from one composed of "spin-right" and "spin-left" electrons, their probability distributions, and hence, their density matrices are identical. The following quote expresses more or less the prevailing opinion on whether or not environmental decoherence solves the measurement problem:

> *Does decoherence solve the measurement problem? Clearly not. What decoherence tells us, is that certain objects appear classical when they are observed. But what is an observation? At some stage, we still have to apply the usual probability rules of quantum theory* (Joos (2000), pp. 1-17).

### 4.2.5 Summary of the Measurement Problem

The previous paragraphs described the many interpretations of quantum mechanics. In addition to the ones discussed above, there are many others (consistent histories, de Broglie–Bohm theory, relational quantum mechanics, transactional interpretation, stochastic mechanics, objective collapse theories, many minds, quantum logic, modal interpretations of quantum theory, and time-symmetric theories). The large number of interpretations suggests that there is no general agreement on what the mathematical formalism of the quantum theory represents.

Viewpoints on this issue fall roughly into three categories: (i) The interpretation problem is primarily a philosophical issue that remains outside the domain of physics, (ii) physics is primarily concerned with whether a theory predicts the correct results of measurements and anything beyond that is outside the domain of physics; (iii) quantum mechanics does not make sense and needs an ontological interpretation or a re-envisioning that does make sense.

As it stands, quantum mechanics consists primarily of a mathematical formalism that correctly predicts a broad range of experimental results. To those whose viewpoint falls within category, (i), the quantum theory in its current form might seem entirely sufficient. Those who fall within category, (ii), are inclined to agree with the category, (i), advocates accept as it pertains to the measurement problem. Since the goal of quantum mechanics is to predict the correct outcome of a measurement, advocates should be able to say what a measurement is. However, category, (iii), advocates believe, despite its successes, that the quantum theory is, on some level, deeply flawed. The mathematics supporting the quantum realm is inconsistent, and its complementarity violates the basic principles of propositional logic. What is observed in the real world are not probabilities but tangible outcomes that have a basis in an external reality that the quantum theory cannot quite touch. Not only does the measurement problem have to be resolved, but the problem of quantum entanglement, as illustrated in the EPR paradox, seemingly violates the principle of local causality (Bell (2011)).

Consistent with wave-particle duality, there is no logical structure, at least one obeying classical propositional logic, which can simultaneously describe all the properties of a quantum system, "*S*." This is often phrased by saying that there are "complementary" propositions, *A* and *B*, that can each describe *S*, but not simultaneously.

Fundamental to propositional logic is the idea of an "atomic statement" or "proposition." A proposition is a statement that "can be the case" or "not be the case" and all other things remain the same (Wittgenstein (1961), § 1.21). "It is raining" is often cited as an example of a proposition. It can be either "true" or "false," but nothing in between. But the "complementary" propositions, *A* and *B*, within the quantum theory are not propositions in the logical sense. In fact, if *A* is true, then *B* is necessarily false, and vice versa. For example, if

$$A \equiv \text{"}I\ get\ an\ exact\ position\ measurement,\text{"}$$

and

$$B \equiv \text{"}I\ get\ an\ exact\ momentum\ measurement,\text{"}$$

then according to the uncertainty principle, if $A$ is true, $B$ is not only false but indefinite. The truth of $B$, in essence, depends on the truth of $A$, which seemingly violates the foundation upon which most mathematical systems are based.

## 4.3 The EPR Paradox

To Einstein, the paradox of wave-particle duality was due to a fundamental lack of knowledge. Nevertheless, thought Einstein, there were reasons, although possibly hidden from view, that would solve the wave-particle paradox and restore determinism to physics. So he set out to show that the quantum theory was, at best, incomplete and, at worse, outright wrong. His efforts, along with collaborators Podolsky and Rosen, culminated in a thought experiment known as the "EPR experiment." But once the experiment was examined and tested, the results would swing the evidence dramatically in favor of the complementarity aspects of the quantum theory and against the idea of "local realism" - the position Einstein favored.

### 4.3.1 The EPR Paper

The now famous paper published in 1935 by Einstein, Podolsky, and Rosen (EPR), *"Can a quantum-mechanical description of physical reality be considered complete,"* discussed what Einstein termed a "spooky action-at-a-distance." Several experiments had revealed that electrons displayed a strange correlation effect. It seemed as though electrons could magically communicate over relatively large distances - distances large enough that the correlations could not be explained by claiming that the electrons could, in some sense, send messages to one another. Einstein knew from his work on the theory of special relativity that the speed of light was finite, and nothing could travel faster. And if nothing could travel faster than light, neither could messages be sent nor information exchanged in less time than it took for light to travel from one place to the next. Thought Einstein, what explanation could be given if the correlations between these quantum objects could not be explained by postulating that information was exchanged between them? The quantum theory didn't provide an answer. In the paper, he argued that something was at work, whether measurable or not, that explained the "spooky action-at-a-distance" correlations. And since the quantum theory did not explain this "something," it was, at best, incomplete and, at worst, outright wrong.

### 4.3.2 The Stern-Gerlach Experiments

Stern and Gerlach (1922) conducted experiments related to the deflection of particles. The experiments showed that electrons and other elementary particles have intrinsic quantum properties. In addition, the experiments demonstrated how, at the quantum level, the act of measuring affects the system being measured.

The experiments involved sending a beam of particles through an inhomogeneous magnetic field and observing particle deflections. The results showed that particles possess an intrinsic angular momentum analogous to the angular momentum of a classically spinning object. This "intrinsic angular momentum" is referred to as "spin."

If the magnetic field is inhomogeneous, the force on one end of the dipole will be slightly greater than the opposing force on the other. The imbalance results in a net force, which deflects the particle's trajectory. If the particles were classical spinning objects, the expected amount of spin would be random and continuous. Each particle would be deflected by a different amount, producing a smooth distribution on a detector screen. Instead, the particles passing through the Stern-Gerlach apparatus deflected either up or down by a specific amount. These results suggested that spin was quantized.

Electrons are spin-1/2 particles. Only two possible spin values measured along any axis ($\pm\hbar/2$) exist. Were electrons classical rotating particles, for example, rotating like a planet around its axis, the individual particles would spin unfeasibly fast. The electron radius measured at 2.8 $fm$ (the classical electron radius) would rotate at $2.3 \times 10^{11}$ $m/s$ - exceeding the speed of light ($2.998 \times 10^{8}$ $m/s$), which, according to the theory of relativity, would be impossible (Tomonaga (1997), p. 35). Instead, spin is a purely quantum mechanical phenomenon.

In the Stern–Gerlach experiments, electrons pass through a device that records spin measurements. In this case, two possible values can occur, "spin-up" ($+\hbar/2$) or "spin-down" ($-\hbar/2$). Spin is a conserved quantity. Hence, if the total spin of all the electrons is zero, and if electron spin is not filtered out in any way, 1/2 of the electrons would be detected spin-up, the other half spin-down. Imagine a filter that prevents, for example, all the spin-down electrons from arriving at the detector. In this case, only spin-up electrons would pass through the filter and be detected. On the other hand, only spin-down electrons will be detected if the filter is rotated by 180° (Harrison (2001)).

Suppose a second filter is placed behind the first. If the second filter is set to filter out spin-down electrons, as does the first filter, the arrangement does not affect the experiment's outcome. All electrons passing through the first filter will pass through the second filter. However, no electrons will be detected if the second filter is rotated by 180°. This is because the first filter will block all the spin-down electrons, and the second will block all spin-up electrons from arriving at the detector. Since electrons are either spin-up or spin-down, no electrons are detected (Harrison (2001)). This experiment demonstrates the quantized nature of spin in electrons. To completely block electron spin, filters need only two discrete settings.

Now suppose a third filter is added. Assume the filters are aligned sequentially and set at 0°, 90°, and 180°. In this case, the first filter only allows electrons with spin in the $+z$-direction through. The second filter filters out spin in the $x$-direction, and the third filter allows only electrons with spin the $-z$-direction through. The electrons that emerge from the 0° filter are all spin-up in the $z$-direction, since all the spin-down electrons have been filtered out. Of all the electrons that pass through the first filter, only $+x$-electorons will pass through the 90° filter. Hence, only 1/2 of the time will electrons be detected by the second filter (90°) (Harrison (2001)), a uniquely quantum mechanical effect. This example illustrates the uncertainty principle and the role of conjugate variables. Conjugate variables cannot be measured simultaneously. If a $+z$-spin measurement is made in the $z$-direction, spin measured in the $x$-direction becomes completely uncertain. There will be a 50% chance of getting a spin-up measurement in the $x$-direction and a 50% chance of getting a spin-down measurement in the $x$-direction.

The first filter lets only spin $+z$-electrons through. But when a $+x$-direction filtering is made, the system completely forgets about the first filtering. This forgetfulness can be understood by noting that no electrons pass through the 180° filter if the second filter is removed. The upshot is that taking a measurement impacts the state of a quantum system (Harrison (2001)).

Stern-Gerlach-type experiments can be designed to examine the correlation effect of spin in electrons. Spin is a conserved quantity which means, barring outside influences like external forces, a system's total spin must remain constant. For example, suppose their total spin is zero before electrons pass through a magnetic field. Then, if ten electrons emerge from the magnetic field as spin-up, there must be ten that emerge spin-down so that the total spin remains zero. Einstein believed that unknown reasons (hidden variables) could explain the spin conservation in electrons and other radiating particles that the quantum theory could not account for.

To get a deeper understanding of the correlations between quantum spinning objects, imagine that two 0° detectors are separated by a distance great enough so that two observers stationed at each detector cannot communicate fast enough to change the results of any measurement. Imagine a radioactive substance originally in a spin-0 state decaying into two electrons emitted nearly simultaneously. One headed toward one of the detectors, and the second toward the other detector. If the decay process is continuous, there will be a continuous stream of electron pairs, each heading toward separate detectors. In this type of decay, 1/2 of the electrons, incident upon each detector, are detected. In addition, if a given detector detects an electron, its entangled partner is not detected by the other detector (Harrison (2001)), which illustrates spin conservation. If one of the electron pairs is spin-up in a given direction, the other must be spin-down for a total spin of zero. The 0° detector detects only spin-up electrons in the $z$-direction. If a given detector detects an electron, it is spin-up. Its entangled partner must be spin-down in that direction.

Suppose one of the detectors is rotated, making it a 180° detector. In this case, 1/2 of the electrons, incident upon each detector are detected. But this time, if one detector detects an electron, the other detector also detects its entangled partner. If a detector did not detect an electron, its entangled partner would go undetected.

Suppose one of the detectors is rotated to 90°, measuring spin in the $+x$-direction. In this case, 1/2 of the electrons, incident upon the 0° detector, are detected while 1/2 of the electrons, incident upon the 90° detector, are detected. But if the 0° detector detects an electron, 1/2 of the time the 90° detector detects its partner and 1/2 of the time, it does not. This is illustrative of the uncertainty principle. If a measurement is obtained for spin in the $+z$-direction, it is impossible to predict what the spin measurement (spin-up or spin-down) will be in the conjugate $x$-direction.

### 4.3.3 The *EPR* Argument

These correlations, or "spooky action-at-a-distance," go by the name, "quantum entanglement," one of the most bizarre, misunderstood, and controversial aspects of quantum mechanics. Two issues are involved: (i) How can quantum entanglement be comprehended as a fundamental feature of our Universe? In other words, quantum entanglement seems to make no sense regarding what is experienced in the observable world; (ii) why, if most quantum states are entangled, are these entanglements never experienced in daily life (Penrose (2005), § 23.6)?

Einstein agreed that "quantum entanglement" made no sense. In the 1935 EPR paper, he argued that the Copenhagen interpretation of quantum mechanics harbored a fundamental flaw (i.e., that conjugate variables, like the position or momentum of a particle, could not be known simultaneously with certainty (wave-particle duality)).

The EPR thought experiment involves two systems that initially interact but then become separated. The position or momentum of one of the systems is measured. Due to the known relationship between the measured value of the first and the second particle, the observer becomes aware of that value in the second particle. A measurement of the conjugate state is then made on the second particle. Once again, the entangled relationship between the two particles reveals that value in the first particle. This outcome would have violated the uncertainty principle, as a single particle's position and momentum could be known simultaneously (Aspec (1999)). For example, two particles, $A$ and $B$, interact briefly and move off in opposite directions. It is possible to measure the exact position of particle, $A$, and the exact momentum of particle, $B$. Here, there is no violation of the uncertainty principle. But because quantum states are entangled, knowing the position of particle, $A$, the exact position of particle, $B$, can be known. Similarly, knowing the exact momentum of particle, $B$, the exact momentum of particle, $A$, can be worked out. The EPR paper argued that it has proved that the exact values of position and momentum can be known simultaneously, directly contradicting the uncertainty principle (Kumar (2009), p. 305).

To see how this argument relates to electron spin, suppose an entangled electron pair is sent toward two spin detectors. One detector is set to measure spin in the $+z$-direction and the other in the $+x$-direction. If an electron is detected as $+z$, its partner is $-z$ in the $z$-direction. Measurement is not required. Correspondingly, if an electron is $+x$, its partner is $-x$ in the $x$ direction. But according to the uncertainty principle, an exact measurement of spin in the $z$-direction precludes knowing the spin in the conjugate $x$-direction. The EPR argument, so it seemed, had beaten the uncertainty principle and showed conclusively that conjugate variables could be known simultaneously.

The EPR thought experiment was entirely at odds with the quantum theory. According to the Copenhagen interpretation, a particle's momentum and position are governed by a single wave function. A position measurement on the particle causes an instantaneous collapse of the wave function. At that point, the wave function ceases to be a factor, making an exact momentum measurement impossible. This was the EPR paradox. Quantum mechanics predicts that both values cannot be known simultaneously, yet the experiment shows they could. Therefore, the quantum mechanical description of physical reality, the EPR argument concluded, was incomplete.

Spin in different directions, like position and momentum, are conjugate variables. In the Stern-Gerlach correlation experiments, if an electron is detected by a 0° detector, it is a $+z$ (spin-up) electron. Once this information is known, its entangled partner is a $-z$ (spin-down) electron. In fact, whatever axis spin is measured along, the spin of the two partner particles is consistently found to be the opposite. This result can only be explained if the particles are linked somehow. Either they were created with a definite (opposite) spin about every axis—a "hidden variable" argument or one electron "feels" which axis the other is having its spin measured along and becomes the opposite about that axis – an "entanglement" argument. Either the electron partner knows that a measurement has taken place, or it has a definite spin already about a second axis. The EPR paper asked how can the second particle "know" to have a precisely defined momentum but an uncertain position. Since this implies that one particle is communicating with the other instantaneously across space faster than light speed and because the quantum theory had no explanation for this "spooky-action-at-a-distance," it must be missing something. This was the crux of the EPR argument (Einstein (1935), pp. 777-780).

The arguments in the EPR paper are based on the principle of locality, which states that physical processes occurring at one place should have no immediate effect on the processes at another location. This is a consequence of special relativity, which maintains that information cannot be transmitted faster than light speed without violating causality. Causality is the relationship between an event (the cause) and a second event (the effect), where the second event is a consequence of the first. It is generally believed that any theory which violates causality would also be internally inconsistent and deeply unsatisfactory. It seems that the usual rules for combining quantum mechanical and classical descriptions violate the principle of locality without violating causality. In the correlation experiments, causality is preserved because there is no way to transmit messages between detectors that allow the manipulation of the measurement results. Therefore, neither the EPR argument nor any quantum experiment demonstrates that faster-than-light signaling is possible. If so, then what is the explanation for the apparent faster-than-light communication between entangled quantum objects? The quantum theory predicts such entangled states in its mathematical formalism but does not explain how they arise.

## 4.4 Bell's Theorem

Almost thirty years passed before any definitive progress toward resolving the EPR paradox would appear. But in 1964, John S. Bell (1987, pp. 195–200) published a paper that contained Bell's Theorem.

### 4.4.1 Bell's Inequality

To understand the significance of Bell's Theorem, it is first necessary to gain an appreciation for the inequality the theorem presupposes. Suppose an object can possess three distinct characteristics, labeled $A, B, C$. Symbolically, Bell's inequality can be written

$$N(A, \sim B) + N(B, \sim C) \geq N(A, \sim C).$$

The above inequality says that the number of objects that possess characteristic, $A$, but not characteristic, $B$ ("$\sim$" stands for "not"), plus the number of objects that possess characteristic, $B$, but not characteristic, $C$, is greater than or equal to the number of objects that possess characteristic, $A$, but not characteristic, $C$.

Bell's inequality is simply an exercise in classical propositional logic. In classical logic, an argument is valid if the premises entail the conclusion. In Bell's inequality, "$N(A, \sim B), N(B, \sim C)$" are the premises and "$N(A, \sim C)$" is the conclusion. The Venn diagram in Figure 4.4.1-1 shows three circles labeled, $A, B, C$, divided into distinct areas marked $a1, a2, a3$ ... The individual circles represent each of the three possible characteristics ($A, B$ or $C$) that an object might possess. The area, $a1$, represents those objects possessing characteristic, $A$, but not characteristics, $B$ or $C$. Area, $a2$, signifies objects possessing characteristics, $A$ and $B$, but not characteristic, $C$. Similarly, area, $a4$, signifies objects possessing all three characteristics.

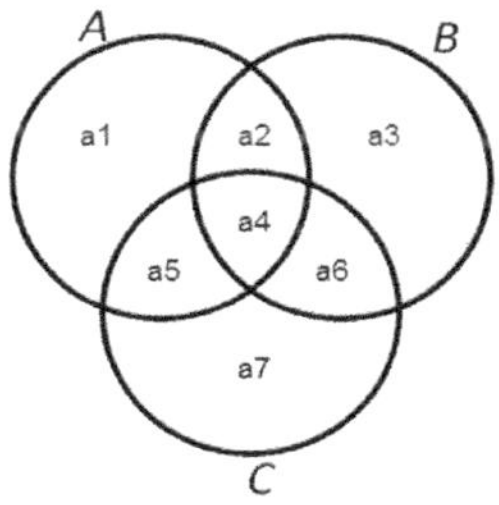


Figure 4.4.1-1.

In terms of Bell's inequality, as it relates to the Venn diagram,

$$N(A, \sim B) \equiv \text{"the number of objects in the areas, } a1 \text{ and } a5."$$

$$N(B, \sim C) \equiv \text{"the number of objects in the areas, } a2 \text{ and } a3."$$

$$N(A, \sim C) \equiv \text{"the number of objects in the areas, } a1 \text{ and } a2."$$

Remembering that Bell's inequality is written

$$N(A, \sim B) + N(B, \sim C) \geq N(A, \sim C),$$

then substituting the Venn diagram values for their equivalent symbols in Bell's inequality gives

$$N(a1 + a5) + N(a2 + a3) \geq N(a1 + a2).$$

The inequality, written in the Venn diagram representation, is equivalent to Bell's inequality. It says symbolically that the number of objects in the area, $a1 + a5$, plus the number of objects in the area, $a2 + a3$, is greater than or equal to the number of objects in the area, $a1 + a2$. Moreover, the inequality is logically true since the right-hand side of the inequality (the conclusion) is entailed in the left-hand side (the premises), which is the definition of a valid logical argument.

David M. Harrison of the University of Toronto gives an excellent example of applying Bell's inequality (Harrison (1999)). In his example, the objects consist of students at a university. The characteristics are:

A. Male.
B. Taller than 5′8″.
C. Blue eyes.

Bell's inequality states that the number of male students who are not taller than 5′8″ plus the number of students (male or female) who are taller than 5′8″ but do not have blue eyes is greater than or equal to the number of male students who do not have blue eyes. According to Bell's inequality and, by extension, propositional logic, this relationship between students is always true.

The truth of Bell's inequality depends on two assumptions: (i) The problem under consideration is conducive to analysis by the methods of classical propositional logic. In other words, the problem's structure is in one-to-one correspondence with classical propositional logic. If this were not the case, then classical propositional logic would be an invalid way of reasoning about a particular problem; (ii) the characteristics of the objects under consideration exist (are true or false) whether observed or not. For example, it can be shown that

$$N(A, \sim B, \sim C) + N(A, B, \sim C) = N(A, \sim C).$$

Note that the above equation's right-hand side equals the right-hand side of Bell's inequality. Even though characteristic, $B$, in this case, is not directly observed, the assumption is that either $\sim B$ or $B$ is true of every object (Harrison (1999)), in this case, every student.

As it turns out, Bell's inequality can be applied to Stern-Gerlach spin correlation experiments. In those experiments, three detectors could be aligned at 0°, 90°, and 180°. If a considerable distance separates two detectors, then two observers stationed at each detector and recording spin measurements cannot communicate in time to influence the measurement results. A radioactive substance originally in a spin-0 state decays into two electrons emitted nearly simultaneously, one electron headed toward one of the detectors, the entangled partner headed toward the other detector. There are only two spin measurements in any direction: "Spin-up" or "spin-down." If one of the electrons is spin-up, the other must be spin-down, ensuring a total spin of zero. The detectors can be constructed so each detector measures spin in a specific direction. If a given detector detects an electron, it is spin-up in that direction. Its partner must be spin-down along the same direction.

It is customary in experiments involving Bell's theorem to divide the detection angles in half, so instead of 0°, 90°, and 180°, the angles are set to 0°, 45°, and 90°. This change allows for testing the uncertainty principle, which involves conjugate variables. According to quantum theory, if an exact measurement is obtained on one of these variables, its conjugate becomes completely uncertain. But in the 0° – 180° detector configuration, if there is detection at 0°, there will be a detection at 180°. Therefore, the 0° – 180° detector configuration is not conjugate. Cutting the detection angles in half overcomes this limitation.

The way Bell's inequality is applied to the electron spin problem is to count the number of electron spin detections in various detector configurations and then examine the results. The characteristics, in this case, are the detection angles:

A. 0°.
B. 45°.
C. 90°.

In the spin correlation experiments, Bell's inequality can be written

$$N(0°, \sim 45°) + N(45°, \sim 90°) \geq N(0°, \sim 90°).$$

The degree numbers in parentheses represent the detector configurations. For example, $N(0°, \sim 45°)$ indicates that the first detector is set to 0° and the second detector to 45° and $N(45°, \sim 90°)$ indicates that the first detector is set to 45° and the second detector is to 90°. In terms of the electron spin, the values inserted into Bell's inequality are the number of detections registered in the 0° detector but not in the 45° detector when the detectors are in the 0° – 45° configuration. Added to that is the number of detections registered in the 45° detector but not in the 90° detector when the detectors are in the 45° – 90° configuration. According to Bell's inequality, that sum should be greater than or equal to the number of detections registered in the 0° detector but not in the 90° detector when the detectors are in the 0° – 90° configuration.

The experiments are usually performed by setting the detectors to the 0° – 45° configuration, then sending a billion electron pairs through the detectors and recording the results. The detectors are then set to the next configuration, and another billion electrons are sent through the detectors, and those results are recorded, and so on. The upshot is that Bell's inequality is violated in the spin correlation experiments. In other words, Bell's inequality does not hold for quantum mechanical systems (Harrison (1999)).

Unsurprisingly, the profundity of this result has come under scrutiny. One argument is that since the two detectors are set at one configuration of angles, and then data is collected for, say, a billion electrons, there is time for the detectors to "know" each other's orientation, although not by any known mechanism. But tests have been conducted where the angles of the filters are set randomly after the electrons have left the source. The results are the same. Bell's inequality is violated, and the predicted quantum correlations are confirmed (Harrison (1999)). Other critiques point out that the correlated pairs emitted go in all directions, so only a tiny fraction of them are measured by the detectors. But experiments using Beryllium atoms have measured almost all of the pairs. The results again confirmed the quantum correlations (Harrison (1999)). There are other objections, but suffice it to say that most physicists believe the quantum correlations have been sufficiently confirmed.

### 4.4.2 Proof by Contradiction

What is Bell's theorem? Bell's theorem is a proof by contradiction, which involves making certain assumptions about the thing to be proved and then showing that the assumptions lead to a contradiction. Hence, at least one of the assumptions must be false. Recall that Bell's inequality is based on two assumptions: (i) Classical propositional logic is a valid way of reasoning about a specific problem; (ii) an object has a definite characteristic, whether or not it is observed.

Bell's theorem applied to the electron spin experiments adds a third assumption (Harrison (1999)):

1. Classical propositional logic is a valid way of reasoning.
2. Electrons have a spin in a given direction, whether measured or not.
3. No information exchange can travel faster than light.

The third assumption covers the EPR paradox and is called the "locality assumption." In fact, if assumptions two and three are valid, it would mean that the uncertainty principle is beaten since (i) nothing travels faster than light and (ii) the spin orientation of electron pairs could be known simultaneously, in direct violation of the uncertainty principle.

Sometimes the three assumptions are revised (Harrison (1999)):

1. Logic is valid.
2. There is a reality separate from its observation.
3. Locality.

The critical insight into the electron spin experiments is that Bell's inequality is violated. This result means that at least one of the assumptions is false, a valid conclusion reached through proof by negation. In his treatment of the subject, Harrison discusses the consequences if each of the assumptions individually turns out to be false while the others remain true (Harrison (1999)). He treats each assumption as if it is independent of the others.

My perspective on Bell's theorem differs from Harrison's. Bell's inequality is an exercise in classical propositional logic. And classical propositional logic is based on the idea of a "proposition," which is fundamental to the subject. A proposition is a statement that "can be the case or not be the case, and everything else remains the same" (Wittgenstein (1961), § 1.12). To be atomic, a proposition must be assigned a truth value (true or false) without affecting the truth value of any other proposition. In this sense, a proposition is local by definition. And it seems reasonable that the statements that make up the assumptions associated with Bell's inequality should be propositions since they are the only meaningful statements within that logical system. For example, $P \rightarrow Q$ ($P$ implies $Q$) is the definition of a compound proposition called an "implication," where $P$ and $Q$ are propositions. If $P$ is true and $Q$ is true, then $P \rightarrow Q$ is true. But, if $P$ is true and $Q$ is false, then $P \rightarrow Q$ is false. The point is that before a truth value can be assigned to the implication, it is necessary to assign a truth value to each proposition that comprises the implication. The same can be said of compound statements like $P \wedge Q$ or $P \vee Q$.

The characteristics in Harrison's "student example" are certainly atomic. For instance, if it is false that a student is taller than 5′8", there is no implication that the student has blue eyes. The same can be said for any combination of the statements associated with the "student example" characteristics. The truth value of a characteristic does not depend on the truth value of one of the other characteristics. But is this true of the assumptions associated with Bell's inequality?

Assuming the assumptions associated with Bell's theorem are propositions, suppose that the third assumption, the "locality" assumption, is false. Could the first assumption (logic is valid) be true simultaneously? In other words, could classical propositional logic be used in some sense to describe a non-local reality? It is tough to understand how it could. To see this, suppose $P$ represents the statement, "the electron is "spin-up," and $Q$ represents the statement, "the electron is not "spin-up," (which means it is spin-down). Considered individually, both $P$ and $Q$ are propositions. Either could be true or false. But, because electron spin comes in entangled states, both statements $P$ and $Q$ could be true of a single electron simultaneously. Such a circumstance is not allowed in classical propositional logic. In a sense, the statement, $Q$, is the same as the statement, $\sim P$. And the statement, $P \wedge \sim P$ ($P$ and $\sim P$), within classical logic is always false. In propositional logic, no single object can have a characteristic and not have that same characteristic simultaneously. In plain language, such statements are contradictory. In essence, the statement, "$P$ and $Q$ are propositions," is the same thing as saying that $P$ and $Q$ are local - their truth values should not depend on the truth values of any other proposition. But if the "locality" assumption is false, then $P$'s truth value determines $Q$'s truth value by necessity. This assertion directly conflicts with the very idea of a proposition. Seeing how the "locality" assumption could be false, and, at the same time, the "logic is valid" assumption remains true is difficult, at least, as it applies to electron spin experiments and, by extension, the quantum theory. The only possible conclusion is that if the "locality" assumption is false, so is the "logic is valid" assumption.

Care must be taken here because many physicists regard the locality assumption as true since it implies that no signal can travel faster than light, a contention virtually all physicists accept. For instance, if a coin is cut in half so that one half contains a head and the other half a tail, and if one of the halves is selected randomly, then the statement, "this half contains a head," is certainly true or false. The only way of knowing if the half selected contains a head is by looking at it. Many physicists regard this as a local phenomenon since if one half of the coin contains a head or a tail, it does not in any way change the fact that the other half of the coin could be a head or a tail. And since there is no direct communication between the two halves of the coin, the phenomenon is deemed local.

But the notion of locality given above is entirely at odds with propositional logic. The statement "this half of the coin contains a head" is true or false, appearing to be a proposition and a meaningful statement within propositional logic. But it is not a valid proposition. To see this, suppose

$$P \equiv \text{"}\mathit{This\ half\ of\ the\ coin\ contains\ a\ head.}\text{"}$$
$$Q \equiv \text{"}\mathit{The\ other\ half\ of\ the\ coin\ contains\ a\ head.}\text{"}$$

Now consider the implication, $P \to Q$. If $P$ is true, commonsense dictates that $P \to Q$ is true, since if $P$ is true, then $Q$ must be false, and if $P$ is false, then $Q$ must be true, and hence, the truth value of $Q$ is determined by the truth value of $P$ (i.e., the truth value of $P$ implies the truth value of $Q$). But within propositional logic, if the antecedent of an implication is true and the consequence false, then the implication is false. But if the antecedent is false and the consequence true, then the implication is true. Hence, there is a direct contradiction between the notion of locality given above and propositional logic. Of course, physicists are free to define locality however they choose. But this does not imply that their notion of locality is consistent with propositional logic. It is not. To be atomic, and hence, a meaningful statement within propositional logic, the truth value of $P$ should not determine the truth value of $Q$. And since if $P$ is true implies that $Q$ is false, the statements, $P$ and $Q$, must be considered statements outside of propositional logic.

Classical propositional logic is primarily based on the law of the excluded middle, which says that propositions are either true or false. No declarative statement within classical propositional logic implies that a proposition is partly true or false. If the assumption "there is a reality separate from its observation" is false, can the assumption "logic is valid" be true? Again, it is hard to see how it could. Consider the statements:

$$P \equiv \text{"}An\ electron\ has\ an\ exact\ position.\text{"}$$
$$Q \equiv \text{"}An\ electron\ has\ an\ exact\ momentum.\text{"}$$

If the assumption "there is a reality separate from its observation" is false, as it applies to the quantum theory, then it would be impossible to assign either $P$ or $Q$ a truth value before an observation is made. Before the observation, the electron is supposedly in a state where it does not have an exact position or an exact momentum. Only probabilities associated with the two propositions can be given. Left in a logical no man's land, only a system of logic not based on the law of the excluded middle might apply. Again, the only reasonable conclusion is that if the assumption, "there is a reality separate from its observation," is false, then so is the assumption, "logic is valid."

On the other hand, once an observation is made, if *P* is true, then *Q* is false and vice versa. At that point, truth values can be assigned, but then the same problem arises as in the previous example, assigning a truth value to one of the propositions determines the truth value of the other. Neither *P* nor *Q*, in this case, is a proposition in the sense of classical propositional logic. If either of the assumptions, two or three, associated with the electron spin experiments were false, it is difficult to see how the "logic is valid" assumption could be true. The truth value of the "logic is valid" assumption depends on the other assumptions' truth value. This violates the essence of a proposition fundamental to classical propositional logic. Therefore, it would seem that

*All three assumptions associated with Bell's theorem are the same.*

Actually, not quite. It is certainly possible that the "locality" assumption is true, while the assumption that "logic is valid" is false. And it is certainly possible the "there is a reality separate from its observation" assumption is true, while the assumption that "logic is valid" is false. But what can be said is if either the "there is a reality separate from its observation" assumption or the "locality" assumption is false, then the assumption, "logic is valid," is false.

Bell's theorem is often cited as a resolution to the EPR paradox. The argument is that since the concept of local realism, favored by Einstein, yields predictions that disagree with those of the quantum theory, and because numerous experiments agree with those predictions, the concept of "local realism" is refuted (Parker (1994, p. 541). It is not hard to see why this is an often-repeated interpretation of Bell's theorem. The objective of applying Bell's theorem is to resolve issues regarding the nature of the physical world. If, by chance, Bell's theorem also has profound implications regarding the language (the logic) physics is written in, that's a bonus but undoubtedly less noteworthy than its "nature of the world" implications.

# Chapter 5: The Philosophical Challenges of Unification

***"The time has come to realise that an interpretation of the universe—even a positivist one—remains unsatisfying unless it covers the interior as well as the exterior of things; mind as well as matter. The true physics is that which will, one day, achieve the inclusion of man in his wholeness in a coherent picture of the world."***

***– Pierre Teilhard de Chardin,***
***The Phenomenon of Man***

## 5.0 Introduction

The limitations on laboratory experimentation have thrust the practitioners of theoretical physics from hypothesizing about things accessible to experimental and observational confirmation into prognostications that reach the edges of speculative philosophy. For example, there is the problem of explaining the nature of dark matter and energy and the issue of the force-carrying bosons, neither of which are directly observable. Quantum theories depend on a "wave function," which admits no basis in reality.

Physicists have attempted to entirely excise metaphysics from physics but without success. But which metaphysical ideas are essential in formulating a valid physical theory, and which are unimportant and why? These are the philosophical challenges associated with finding a unifying theory for physics and are the topics of this Chapter.

## 5.1 Two Philosophical Questions

Two unresolved philosophical questions play a significant role in unifying physics. The first is the "mind-body problem." Is there such a thing as a "mind," and if so, is the mind independent of the body? Physicists often state the problem in the following way: "What is consciousness," and what role should it play in physics?

The second unresolved question involves whether an external world exists independently of our perception of it. Again, this question reflects the antithesis of the realist/idealist worldviews.

To philosophers, and I should think to most physicists too, it is essential to know whether the observations and the measurements extracted from them reflect the true nature of the object being observed. Yet, interestingly, some physicists take an agnostic view toward this question, claiming that theories should only predict the correct outcome of a measurement. They claim that all questions beyond this are not physics. But, then, what does a measurement represent exactly?

"Positivism" asserts that the only authentic world knowledge should involve sense experience and positive verification. The logical positivists of the early twentieth century concluded that pure abstractions could tell us nothing about an external world. According to this view, the world consists of facts arranged into certain relationships called "states of affairs." The facts and states of affairs make the world what it is. Facts and states of affairs are of a particular kind. To the positivist, a man could point to an object and say "cat." This act assigned a sound (symbol) to a particular object in space and time. But the symbol represented a particular object and a particular state of affairs. For example, if the man thought he had formed the abstract notion of a "cat," he was mistaken.

If an ordinary citizen were asked, "What is a cat?" the reaction would likely be one of surprise. But a positivist would point out that even though we could point to dozens of objects and say, "This is a cat," we could not, with certainty, recognize the next cat if we saw it. Our notion of a "cat" is purely an idea. There would be no physical object that exactly matched our idea of a cat. Cats are objects characterized by a particular set of sense data, which cannot be further analyzed beyond sense experience. We, as humans, are inclined to associate a particular array of sense data with the word, "cat," which represents what we see with our eyes, etc. The positivist, while admitting the existence of an external world, would deny there is a one-to-one correspondence between our abstract notions of an external object and the object itself.

The positivist viewpoint has undergone several modifications that attempt to clarify what "positive verification" means. It is safe to say the positivist tenant of direct "positive verification" is not strictly adhered to today. Its influence, however, carries considerable weight within the physics community.

What does modern physics say about the two questions just posed? First, it has much to say about the "Is there an external world?" question. The modern physics answer is one of the most controversial of current theoretical thought. The theory of relativity answers the question in the affirmative, while the quantum theory gives a somewhat qualified negative answer. Most quantum theorists would accept that an external world exists but would add that our attempts at observing it are modified by observation. And, some would argue, modified by consciousness. According to this view, it cannot be genuinely claimed that there is an independent external world separate from our observation of it. Even if there is, nothing definitive can be said about it. Any attempt at penetrating the world of the wave function to gain a deeper understanding of the reality of that world has generally failed.

Unlike relativity, which assigns an objective reality to external objects, whether observed or not, quantum mechanics replaces local realism with a kind of positivism. The quantum theory regards the unobserved world as unreachable except to give probabilities to outcomes. Nothing can be said about the realism of the world until an observation. One of the most perplexing mysteries in theoretical physics is how these two theories can be reconciled into a sensible theory that resolves the dichotomies in each of their philosophical perspectives.

As to the first question, at least in its present form, current theoretical thought in physics has little to say. Nevertheless, physicists have not ignored the question. On the contrary, there does not appear to be one approach or angle of attacking the question that has gained general acceptance. A "conscious observer" plays a significant role in the measurement problem for those physicists who maintain that, in some sense, consciousness collapses the wave function.

However, past philosophers had plenty to say about the first and second questions. And it will be instructive to examine what they had to say, if for no other reason, but to understand better the two issues from the varied contemplations of some of the finest minds of the past centuries.

### 5.1.1 René Descartes (1596 – 1650)

Descartes thought that knowledge gained through sense experience was unreliable. Only through thought, wholly divorced from the influence of sense, could anyone truly understand anything. To Descartes, the mind was a powerful, independent agent, the path to true knowledge. He believed that the body worked like a machine and had the material properties of extension and motion reflected in the laws of physics. On the other hand, the mind was non-material, lacked extension and motion, and was not subject to the laws of physics.

Descartes believed that the mind was distinct from the body. Here, Descartes means a distinction between two or more substances. A fundamental distinction is perceived when one substance can be clearly and distinctly understood without knowing anything about the other. In arguing for a fundamental distinction between mind and body, Descartes claimed that (i) the mind is a substance, (ii) it can be distinctly understood without any other substance, including a body, and (iii) minds or souls exist without bodies (Rodreguez-Peretra (2008)).

Descartes elaborates on the differences between material and non-material substances by claiming that the nature of the body or extension is divisible into parts. In contrast, the nature of the mind is "something quite simple and complete," in a word, "indivisible." Therefore, the mind and body cannot have the same nature, for if this were true, then the same thing would be both divisible and indivisible, which, to Descartes, was impossible. Hence, mind and body must have two completely different natures for each to be understood by itself without knowing anything about the other (Rodreguez-Peretra (2008)).

The origins of the mind-body problem trace back to the meditations of Descartes and his conclusion that the mind and body are distinct substances. But Descartes ran into difficulty. Supposing the mind is an entirely non-material thing without any extension whatsoever, and the body is an entirely material thing without any thinking, then each substance can have only its particular modes. For example, the mind has modes of understanding, will, and, in some sense, sensation, while the body has modes of size, shape, motion, and quantity. But if that is the case, why can the "will" move the body, for example, when a question in a student's mind causes the raising of the arm, and certain motions in the body cause the mind to have sensations? In other words, how can two substances with completely different natures causally interact? Descartes could not provide a convincing explanation for this question (Rodreguez-Peretra (2008)).

Descartes thought that deductive reasoning was the only path to an accurate and complete understanding of nature. And this could only be achieved through a purely mental discipline free from the deceptions of sense experience. Later, using the deductive method he endorsed, Descartes gave an ontological proof of the existence of God—still, the proof was embodied in the premise of the argument. Descartes's argument was circular. But once he proved the existence of God, he could have faith that God would not deceive him, and through the power and understanding God gave him, Descartes believed he could gain an understanding of the true nature of the external world (Descartes (1968)).

Descartes also responded to the skepticism concerning the existence of an external world by arguing that he does not will sensory perceptions. Instead, they are external to his senses, providing evidence of an external world outside his mind. Furthermore, since the mind was non-material, it could distinguish between itself and substances different from itself. This evidence showed that he could not be deceived that the external world contained material things different from his mind. He also maintained that the cause of the ideas in his mind were the bodies that existed in the external world (Descartes (1968)).

Descartes was a rationalist and a realist. He believed in the independence of mind and body and the independent existence of an external world. But he failed to show that the only true path to a sensible theory of nature was gained through deductive reasoning. Later, theoretical thinkers more inclined to modern scientific philosophy (positivism) insisted that metaphysical statements of the kind that Descartes advocated should be accompanied by verification through experiment and observation.

### 5.1.2 Benedict DeSpinoza (1632 - 1677)

Spinoza thought that Descartes had erred in believing that the mind and the body were different substances. Instead, he believed that the mind and the body acted as a single unit but were, in some sense, independent. Spinoza is considered a monist, but his monism is unique. Usually, monistic theories purport that all existence is made of the same "stuff," material or non-material, whatever the case. But Spinoza's system is different, and he goes to great lengths to explain how the mind and body can be considered a single unit yet maintain a fundamentally independent relationship. It will be beneficial to discuss in some detail the nuances of his system.

Reading Spinoza's monumental work, "*The ethics*," the first few words give an insight into how he planned to treat the mind-body problem. Part One, Definition II reads (De Spinoza (1677)):

> *"That thing is called finite in its own kind (in suo genere) which can be limited by another thing of the same nature. For example, a body is called finite because we always conceive another which is greater. So, a thought is limited by another thought; but a body is not limited by a thought, nor a thought by a body."*

Spinoza reckoned thoughts amongst things non-material and bodies amongst material things. He believed in an independent external world. In this, he agreed with Descartes: Book II, axiom II:

> *"Modes of thought, such as love, desire, or the emotions of the mind, by whatever name they may be called, do not exist unless in the same individual exists the idea of the thing loved, desired, etc."*

Here Spinoza associates an idea with the object of that idea. Presumably, these objects, whether real or imagined, are things perceived as external to the mind. They are things extended in space and time, in other words, material objects. Interestingly, there is no room in Spinoza's system for statements like, "I desire to be happy." Happiness, thought Spinoza, is not a state of mind but arises from the mind's perception of the disposition of the human body. If the body's power of acting, and by the power of acting, Spinoza means the power of existing, is perceived by the mind as enhanced, the mind will rejoice. If, on the other hand, the mind perceives that the body's power to act is hindered, the mind will suffer. Spinoza makes this mind-body connection very clear in Proposition XIII:

> *"The object of the idea constituting the human mind is a body, or a certain mode of extension actually existing, and nothing else."*

Spinoza believed in the existence of a single substance, which he called "God." But Spinoza's God did not stand on the outside looking in as maintained in some religious systems. His "substance" was reflected in the essence of all existing things. Whether it was substance thinking or substance extended, it was the same substance. His stance on this question opposed Descartes', where extended (material) things were subject to the laws of physics, but thoughts (non-material) were not. While Spinoza agreed with Descartes that thoughts were non-material and extended things were material, he believed the same laws governed both extended bodies and non-material thoughts. This stance he made clear in Proposition VII of Book II:

> *"The order and connection of ideas is the same as the order and connection of things."*

This meant that if there were a set of equations representing the laws of physics of the entire material world, those same equations would also describe the laws of thought, even though the material (things) and the non-material (mind) are entirely separate.

This idea is further expressed in proposition XX, part II:

> *"There exists in God the idea or knowledge of the human mind, which follows in Him and is related to Him in the same way as the idea or knowledge of the human body."*

This position opposes the positivist view in which abstract concepts such as "I can think of a cat" are generally considered meaningless. A positivist might concede that God, if such a being exists, might be capable of thinking abstract thoughts of the kind Spinoza describes, but humans certainly could not. But Spinoza believed that individuals could form abstract ideas, if only in a somewhat indirect fashion.

How did Spinoza view the connection between the mind and the body, and what did he mean by the mind and body functioning as a unit? He addresses the first of these questions in Proposition XXIII, part II:

> *"The mind does not know itself except in so far as it perceives the ideas of the modifications of the human body."*

In Spinoza's system, the human mind cannot know anything except through the disposition of the human body. So how is it possible to form an adequate concept of the nature of the external world? His views on this are reflected in Proposition XXV, part II:

> *"The idea of each modification of the human body does not involve an adequate knowledge of an external body."*

And again, in Proposition XXVI:

> *"The human mind perceives no external body as actually existing unless through the ideas of the modifications of its own body."*

Here Spinoza addresses one of the most critical aspects of the realist worldview - that an external world exists independent of our perception of it. Any realist must accept as a possibility that the perceptions of the mind presumably caused by the presence of an external object do not necessarily reflect the nature of that object. Spinoza concedes that all bodies have some things in common, and their natures agree to the extent of that commonality. But he also concludes bodies disagree to some extent. So the knowledge of the nature of an external object cannot be gained through a complete knowledge of the modifications of the human body, which, in his system, is the only thing that the human mind can perceive.

It remains to explain Spinoza's position regarding man's ability to understand the nature of the external world and explain how he thought this knowledge was acquired. Here Spinoza gives a somewhat unconvincing and difficult-to-follow argument. In Proposition XXXVIII, part II, he declares:

> *"Those things which are common to everything, and which are equally in the part and in the whole, can only be adequately conceived."*

Here Spinoza gives hope that a true idea of substance can be gained in finding things that, in some sense, can be conceived as "equally in the part and as in the whole" (whatever this means). It should be clear by now that, in Spinoza's system, a true idea must agree with the actual nature of the thing being considered. So, the something, *X*, conceived through thought, and the something, *X*, external to the mind must have the same structure. This view reflects Spinoza's conception of a substance. Apparently, the knowledge gained is not the nature of the external objects in the world but the general nature of substance (i.e., God). Proposition XXXIX, Book II:

> *"There will exist in the human mind an adequate idea of that which is common and proper to the human body, and to any external bodies by which the human body is generally affected – of that which equally in the part of each of these external bodies and in the whole is common and proper."*

Even though Spinoza's true (adequate) idea is an idea of no individual thing, the human mind is capable of forming an abstract (general) idea of the substance reflected in the nature of what is common and proper to the human body which, in turn, is reflected as common to all external bodies, because, as with everything in Spinoza's system, all things are derived from a single substance. This knowledge Spinoza called "scientific intuition," the ability to grasp the entire nature of substance all at once - in a flash of purely luminous insight.

### 5.1.3 A Summary of Rationalism

By whatever name they called it, rationalist philosophers, to one extent or another, believed that man was endowed with a higher ability that surpassed all the other inhabitants of the Earth. This ability is typically called "reason." For example, Blanchard ((1964), § 2) describes reason this way:

> *"When we say that man is a rational animal, then, we seem to imply that he can command ideas independently of sense, that he can abstract, that he can infer explicitly, and that he can sit in judgment on himself. The highest of animals can do none of these things. The stupidest man, if not a pathological case, can in some measure do them all."*

### 5.1.4 David Hume (1711 - 1776)

No sooner had the ink dried on the pages of the rationalist's tablets was their optimism for man's future, confident insights, and sure judgments tersely and skillfully challenged by the skepticism of their empirically leaning counterparts. While the empirical philosophers did not always agree with one another, the one thing they did agree upon was that human beings possessed no such abstracting ability of the sort claimed by the rationalists. No foolproof general approach could lead to the truth.

David Hume was particularly adamant that the only power the mind possessed was the power of associations. It was not reason but custom and the repetition of associating one impression with another that determined our beliefs. Those beliefs could only be probabilistic. Our beliefs were only likely, not certain. Hume classifies the perceptions of the mind into two distinct kinds: "Impressions" (sensations, passions, and emotions) and "ideas." But he regards ideas as simply faint impressions. All impressions are sense impressions. How he concludes that ideas and impressions are of two distinct kinds is unclear since the difference is only of degree. But Hume's contemplations foreshadowed the "scientific philosophy," where observation and experience play fundamental roles and metaphysical (meta: meaning beyond) ideas which cannot be linked to a particular impression is regarded as meaningless. If Spinoza thought he could know a single substance through some obscure mental abstracting ability, he was mistaken. There was nothing but particular substances reflected in the impressions of sense (Hume (1968), § XIL).

Hume's philosophy greatly influenced the twentieth-century logical positivists. They maintained that the only meaningful statements about the world were those verified through sense, logical tautologies (truth statements), or contradictions (logically false statements). Emotive statements, expressing attitudes, could say nothing about the external world. Likewise, metaphysical statements, like the ones proffered by the rationalists, were meaningless musings that, like emotive statements, could tell us nothing about an external world.

### 5.1.5 John Locke (1632 - 1704)

It was John Locke who struck the first skeptical blow. By the time Locke wrote his *"An essay concerning human understanding,"* Newton had finished his monumental work. Nevertheless, Locke was profoundly influenced by Newton. He believed beyond doubt that the world was mechanical, acting like a giant machine, and having a fundamentally corpuscular (atomic) nature - made from a conglomeration of insensible particles acting on one another by impulse, just as Newton had explained (Warnock (1963), § III).

Like the rationalists, he accepted the notion of an independent external world. Perceptions, he thought, resulted from atoms acting mechanically on the sense organs which sent impulses to the brain.

But when the impulses reached the brain, the material interactions produced a non-material "idea." And it was non-material ideas of which human awareness consisted (Warnock (1963), § III).

Here is where Locke parted ways with the rationalists. Although he endorsed the idea of an independently existing external world, he maintained that man was irreversibly cut off from it. Direct awareness was given to only our ideas, not the objects existing in the external world.

But were the ideas held in our mind faithful representations of the external objects that are taken to be their cause? Locke maintained that some qualities associated with external objects were primary, some secondary. The primary qualities (extension, moving, having mass/weight) really are "in" external objects. Still, owing to the nature of our sensibility, secondary qualities (color, odor, etc.) are merely appearances, figments of the imagination that are not really "in" extended objects (Warnock (1963), § III).

Upon this point, his critics pounced. What criteria had Locke used to distinguish primary from secondary qualities? Locke admitted that what could be perceived directly was the ideas in our heads, not the external objects that supposedly caused those ideas. In Locke's system, non-material ideas were the only things we could be aware of. What grounds did he have for supposing that something was included in those ideas that made some sensible qualities real while others were imaginary? Locke never gave a satisfactory retort to this objection (Warnock (1963), § III).

Locke's position on the question of substance was much like that of Descartes. He firmly believed in material as well as non-material substances. There must be, thought Locke, something to which the qualities of objects belonged, for it cannot intelligently be supposed that the qualities of objects exist in their own right. Therefore, there must be something for them to be qualities of. But this is where the similarities between Locke's view and Descartes' ended. In Locke's mind, there was no way of saying what substance was. For, to anything that pertained to substance was to assign a quality to it unavoidably. And that got no closer to the truth of what substance entailed (Warnock (1963), § III). This position reflected the words that Locke will forever be remembered: Substance is "something, I know not what."

His position on substance pushed Locke into further difficulties. He wished to claim that there were two kinds of substance: "Material" (extended world) and "non-material" (ideas). But if substance was "something, I know not what," on what grounds would Locke be entitled to claim that there were two? In addition, from a theory of knowledge standpoint, Locke's forecast portends a dreary outcome. Of what, then, could we know through contemplating our ideas? If all that we can perceive is our ideas and those ideas, as Locke admitted, were not faithful representations of an external world, neither were they faithful reflections of an underlying substance, then what were they? How could anything be known at all? Locke could not answer this question (Warnock (1963), § III).

### 5.1.6 George Berkeley (1685 – 1753)

Berkeley found Locke's system not only distasteful but also ridiculous and an affront to common sense. Berkeley was devoutly religious, and in Locke's system, he saw an intolerable pretext for religious skepticism. Locke's system was founded on the Newtonian idea of a physical universe operating mechanically like a vast machine. But where in this system did God come in? Locke had assumed that the universe was put into existence and got going by a supreme deity. But Berkeley rightly realized that this assumption could be quickly challenged. What if the universe was eternal? Without a beginning, there would be no room for a creator. In Berkeley's mind, such a scenario was unthinkable (Warnock (1963), § IV).

Berkeley was keenly aware of the struggles Locke encountered trying to explain the differences between the primary and the secondary qualities of matter and his torturous grappling with the idea of two distinct substances, the existence of which Locke had little right to claim (Warnock (1963), § IV).

Berkeley found it absurd that the "visible beauty of creation" is no more than a "false imaginary glare" that our gardens' flowers have no color or scent. He realized that Locke had no basis on which to claim that the qualities of matter differed in any way. Locke's insistence on the existence of an external world of which nothing could be known directly left him no grounds for asserting the existence of the primary and secondary qualities of matter. Upon what would this assertion be based? Moreover, Locke's tenant that the only perceivable things are our ideas, yet supposing those ideas reflect qualities and are connected to an external "something, I know not what," in Berkeley's view, was outrageous, bordering on blatant insanity (Warnock (1963), § IV).

So, what was Berkeley's solution? It was simply to deny the existence of matter. Berkeley believed he could wipe out Locke's tortuous struggles in one blow. It removed at once Locke's bumbling dualistic treatment of substance. A material substance did not exist, so the difficulties associated with Locke's groundless insistence that two different types of substance existed vanished. There existed only a non-material spiritual substance and no other. All of Locke's excruciatingly painful grappling with the absurd hypothesis of the existence of an inaccessible and unknowable external world was wiped away. It simply did not exist. The houses, the trees, and the birds would be associated directly with our ideas and did really exist. Even better, the need for contrasting our ideas with a baffling array of inaccessible and inexplicable causes disappeared (Warnock (1963), § V).

Still, it could be objected that Berkeley's explanation left no account for the existence of our ideas. If it wasn't external objects that caused our ideas, where did they come from? Berkeley's explanation was a truly incorporeal, active substance or spirit whose force of will implanted ideas in our minds in an orderly, systematic, and rational manner (Warnock (1963), § V).

A further objection was that Berkeley's system left no room for the existence of evil. If God caused our ideas and God was good, why was there evil in the world - a condition Berkeley himself was incessantly preoccupied with? Berkeley replied that God provided the orderly connection and arrangement of our world of ideas as experienced. Our relationship with that orderly world of ideas was left up to us (Warnock (1963), § V).

Finally, doesn't this system require that ideas exist only when those ideas are perceived? Yes, Berkeley admitted, but he saw no difficulty in this. More than any other, this insight is undoubtedly the one for which he will be most remembered. In this, Berkeley became prophetic. The scientists of the twentieth century found this insight remarkably useful in explaining the emerging field of quantum mechanics (Warnock (1963), § V).

So, what was Berkeley's position on abstract ideas? Early on, he riled vehemently against them. Abstract ideas were, he thought, the primary reason that the philosophers of the Enlightenment had gone wrong, failed to clear up old errors and introduced new ones, filled the public consciousness with all sorts of unnecessary notions of "something, I know not what" that not only was an affront to common sense but laid thicker the learned dust of the middle ages. His distrust of abstract notions was one of the primary reasons he challenged the ruinous and obscure ramblings of the philosophers of the day. Berkeley thought the answers to all these questions were simple and luminously clear (Warnock (1963), § VII).

But later, Berkeley realized he could not square his position on abstract ideas with his philosophy. He wished to argue that the only things directly perceived were our ideas, which were spiritual in nature. But what was the mechanism that allowed the conclusion that ideas were non-material? Indeed, this perception was not gained through the ideas themselves, the content of which was concrete sense experience. This concept left no room for a deity. Berkeley argued, rather weakly, that in addition to perceptions, there are notions of spiritual things. He argued this without further clarifying what he meant by these "notions" (Warnock (1963), § VII). The notions seemed like abstract ideas, the very ideas for which the rationalists had been condemned. And if it wasn't a perception or an abstract idea, then what was it? This is the weakest of Berkeley's arguments, and he never adequately addressed the issue.

Finally, what account can Berkeley give of the physical sciences? The scientists of the day gave a rather convincing account of how the world acted like a finely tuned machine, a world of atoms acting impulsively on the sense organs providing an awareness of an intelligently inspired external world. An external world, in Berkeley's view, that did not exist. Since a physical substance did not exist, there was nothing in Berkeley's view for the so-called "laws of nature" to be true of. Was the truth of an independently existing external world a figment of the scientific imagination? Berkeley's answer was yes. The scientific notion of an independently existing external world was fiction, but it was a useful fiction. If the scientists found it necessary to invent cleverly devised abstractions to explain in general terms the nature of how our ideas worked, so be it. Berkeley had no objection. But these abstractions were only useful, not true (Warnock (1963), § VI).

Berkeley's insights predate those of some physicists today who are inclined to argue just as he did that today's physical theories are not factual truths but are merely mathematical and predictive conveniences. Not surprisingly, this view encounters stiff resistance. But, if it is not "truth" the physical scientist is after, then what is it (Warnock (1963), § X)?

### 5.1.7 Immanuel Kant (1724–1804)

Kant wrote his Critique of Pure Reason toward the end of the Enlightenment, which was in a state of crisis by then. The Enlightenment had seen the spectacular achievement of Newton, enjoyed a clear break from the oppressive church-dominated darkness of the Middle Ages, and uncovered a new power embodied in human reason. This new confidence in reason brought with it the hope that we could break free from traditional political or religious authorities and dart into a new era of self-government guided by reason. Why should political or religious authorities tell us how to live or what to believe if each of us can figure these things out for ourselves? Nevertheless, Kant was committed to preserving the rationalist dream. To Kant, reason embodied the true path to freedom (Rohlf (2010), § 2.1).

One of the inspirations of the Enlightenment was the new physics. But the new physics was mechanistic - a sea of impulsively guided atoms and bodies governed by causal laws. So where did freedom, a soul, or anything but matter in motion come in? Freedom was necessary for choosing right over wrong because who can be held accountable otherwise? And where was there room for belief in a soul that can survive death and resurrect in an afterlife? Modern science, the source of so much optimism, threatened to undermine the traditional moral and religious beliefs that free rational thought was expected to support. This conundrum reflected the central intellectual crisis of the Enlightenment (Rohlf (2010), § 2.1).

The Critique of Pure Reason was Kant's response to this crisis. The authority of reason was in question, and Kant's main goal was to show through a critique of reason by reason that all essential human interests were mutually consistent. In addition, he attempted to show that reason really does deserve the sovereignty attributed to it by the Enlightenment (Rohlf (2010), § 2.2).

Kant understood his predecessors' struggles to answer the fundamental questions of the time and the associated disputes that naturally arose from those attempts. Against the empiricists, particularly Berkeley, he argued that if all our perceptions were our own ideas, a contention upon which both Locke and Berkeley agreed, how could an awareness of self be accomplished? Here Kant attempts to reestablish the existence of an independent external world - a claim Berkeley flatly rejected. Kant argued that if the only objects of our awareness are our own ideas, it would be impossible to distinguish those ideas from those of objects that have nothing to do with us. Kant firmly maintained that the only way I could distinguish "myself" from things "not myself" is for there to be an external world that would allow me to compare "myself" to things "not myself." Kant's methodological innovation was to employ a "transcendental argument." Kant argues that "there are objects that exist in space and time outside of me, which is necessary for establishing one's own existence. He maintained it would not be possible to be aware of myself as existing without presupposing the existence of something permanent outside of me to distinguish "myself" from. Since I know myself as existing, something is permanent outside of me (Rohlf (2010), § 4.1).

He charged the empiricists with hypocrisy. If Berkeley could ontologically state that matter did not exist, then that claim, according to Kant, necessarily required exercising judgment, and judgment did not come from sense experience alone. Judgment, thought Kant, was primarily a rational endeavor. Therefore, we could judge because we were rational. The empiricists, Kant argued, had committed the ultimate logical sin. They had used reason, in the sense that the rationalists had understood the word, to argue that man had no such reasoning ability (Rohlf (2010), § 4.1-2).

Against the rationalists, Kant argues that *apriori* ideas like "God is a perfect being" are not possible. Kant completely rejects the notion that such propositions are etched on the fabric of the mind. Statements like "God is a perfect being" are *apriori* statements about things as they are in themselves. Consequently, the statement purports to tell us something about God as he is in himself. Such statements, Kant maintains, are impossible because the rational judgments made by the mind can only apply to the objects existing in the external world, revealed in sense experience. The mind lacks content until interaction with the world actuates these formal constraints. The mind possesses *apriori* templates for judgments, not *apriori* judgments completely apart from sense experience (Rohlf (2010), § 4.1-2).

Kant held that it was impossible to know the objects in the external world as they are in themselves. Instead, because we were rational, we necessarily interjected upon those external objects a specific order and connection that was not necessarily inherent in them. In essence, reason allows us to create an orderly structure out of sense experiences (Rohlf (2010), § 4.1-2).

Kant claims that the knower's mind actively contributes to how external objects are experienced, which is fundamental to his "transcendental idealism." His arguments are aimed at limiting knowledge. The rationalists believed that metaphysical knowledge about God, souls, substances, etc. was genuine. But Kant argues that knowledge cannot go beyond the realm of sense experience (Rohlf (2010), § 2). Kant's arguments are not unlike those in quantum mechanics (i.e., that the external world is unknowable until an observation). But quantum mechanists generally accept that the observation is real and not "cooked" by the rationality of the human mind, as Kant suggested.

Kant argues that objects cannot be experienced without being represented spatially. It is impossible to grasp an object as an object unless we can delineate the region of space it occupies. Without a spatial representation, sensations are undifferentiated and not ascribable to a particular object. Time is also a necessary intuition. The idea of "time" cannot be gathered from experience because succession and simultaneity of objects, both things that would indicate the passage of time, would be unrepresentable if the capacity to represent objects in time was not an inherent human ability. In other words, it is impossible to experience objects not in time and space. Space and time are templates and cannot be perceived directly. Any consciousness like ours must apprehend objects as occupying a region of space and persisting for some time. Kant argues that the understanding must provide the concepts and the rules for identifying what is common or universal. By applying concepts, the understanding takes the particulars given in sensation and identifies what is common and general about them. The understanding allows thinking in abstract terms to form general notions of houses, cats, and trees (Rohlf (2010), § 4.2). Here Kant reestablishes the ability of man to form abstract or general ideas, something the empiricists, particularly Hume, flatly rejected.

Kant enumerated his *apriori* contributions to experience in a "table of categories." His list of *apriori* concepts made all empirical judgments possible. These concepts cannot be experienced directly; but are the manifestations that particular judgments of objects take (Rohlf (2010), § 4.2).

What did Kant say specifically about the two fundamental questions that are the primary topic of this discussion (i.e., does an external world exist independently of our perception of it? And is there an active mind independent of the body?)

Kant answers both questions in the affirmative. An external world exists because I can distinguish myself from it. The mind exists because it possesses its own attributes manifested through *apriori* concepts (embodied in the categories) that were independent of pure sensory experience.

But to the aspirations of his rationalist predecessors, who believed that reason could provide a faithful knowledge and representation of existence itself, Kant levied a severe limitation. In Kant's view, the answers to purely metaphysical questions such as "Does God exist?" were beyond reach. Reason could only be applied to the domain of sense experience. Moreover, a faithful knowledge of an external world as it exists in itself was too out of reach. The rational human mind imposes a particular order and connection on the things in the external world. We could make sense of the external world because our rational mind imposed a specific order and connection on things, not because the things in the external world possessed that order and connection. In this, Kant sympathized with the "good Berkeley." The real world was not the vast impersonal machine described by Newtonian physics. On the contrary, our *apriori* template of space and time projected a mechanistic order and connection onto the external world.

Kant's system articulated four types of judgments. There were "*apriori*" judgments made independently of sense and "post-priori" judgments that were not. Those two categories were further divided into "analytic" and "synthetic" judgments. Analytical judgments were those which did not supply new information about a subject. "That tree has a trunk" is an example of a "post-priori" analytical judgment. Presumably, the knowledge of "that tree" could only come by experiencing it. But, having a trunk is a necessary condition for being a tree. The judgment is analytic when no additional information is provided about the subject. But "that tree is 50 feet tall" is an example of a "post-priori" synthetic judgment. It is not necessarily in the nature of a tree to be 50 feet tall. Since being 50 feet tall is new information not inherent in the subject, the judgment is synthetic (Blanshard (1964), § 32-39).

The foundation of Kant's system rested mainly on "*apriori*" synthetic judgments. Kant believed this type of judgment was manifested in mathematics and, in particular, geometry. Thought Kant, 5 + 7 = 12 was an example of an "*apriori*" synthetic judgment. To produce 12 (in this case, 12 is the subject), knowledge of addition is required. And knowing how to add was not part of being the number, 12. It was something new not contained in the definition of 12. By that argument, the judgment is synthetic.

Kant felt that we could rationally project objects in space and time, giving order to the external world, by making "*apriori*" synthetic judgments. Those judgments revealed themselves in Euclidean geometry applied to Newtonian physics. In this, Kant had the utmost confidence. Euclidean geometry was etched into the fabric of the human mind as surely as 2 + 2 = 4. Euclidean geometry allowed for making sense of the external world (Blanshard (1964), § 2.32-2.39).

Kant's critics pounced upon these last two ideas, particularly the logical positivists. Kant believed that our reason gave us order and connection to the physical world. Reason became the rose-colored glasses through which we formulated the ideas of how the world worked. The Universe was orderly, not because that order was inherent in the things in Universe, but because the human mind had "cooked" the result. Our reasoning abilities forced upon Nature an orderly structure whose foundations resided in Euclidean geometry. Nature wasn't the source of the orderly connections we saw. Our *apriori* synthetic judgments embodied in Euclidean geometry gave meaning to the world. Kant was as much a believer in Newtonian mechanics as was Locke. Newton had employed Euclidean geometry to describe the world's workings, and it never seemed to have failed him. To Kant, Euclidean geometry was the necessary connection between pure mathematics and physics. It allowed man to apply mathematics to the description of the physical world (Blanshard (1964), § 3.16-3.18).

But the positivists of the early twentieth century were well versed in mathematics, and by then, non-Euclidean geometries had arrived and been thoroughly researched. Einstein employed non-Euclidean geometry to produce his theory of gravity. These events showed conclusively that Euclidean geometry was not the necessary prescription for describing the world that Kant thought it was. It showed irrefutably that Kant had severely erred. Not only had Kant erred about the nature of Euclidean geometry, concluded the positivists, but he was also wrong about the existence of *apriori* synthetic judgments. There were no such judgments. According to the positivists, statements like "5 + 7 = 12" that Kant thought were synthetic were analytic. In fact, they concluded that all mathematical and logical statements were *apriori* analytic (Blanshard (1964), § 3.18-3.19).

Such a view was a significant departure. It cut off the human connection to the external world that Kant had worked so hard to reestablish against the suspicions of the empiricists. Gone were "*apriori*" synthetic judgments embodied in logic and mathematics that were supposed to make the world a rational and intelligible place. Instead, logic and mathematics became self-contained tautologies - self-consistent systems of pure thought. Lost was any connection to the physical world. To the positivists, logic and mathematics became valid ways of reasoning, but those ways said nothing about the external world (Blanshard (1964), § 3.18-3.19).

Unlike Berkeley, positivists believed in an external world independent of our perception of it but rejected the idea of any connection between mind and body. Instead, logical and mathematical statements were purely rational, with no connection to the external world. Statements like "this apple is red" were considered propositions that could be either true or false. The only way to establish the truth or falsity of such a statement was to verify it through sense experience. Logic and mathematics could provide a self-consistent way of using language but said nothing about the external world.

### 5.1.8 Monistic Idealism

In recent years, several popular books have promoted the idea that the spiritualism of Eastern Mysticism is hidden beneath the formalism of quantum mechanics. These works seek to bridge the conceptual chasms that divide science from religion. Roughly speaking, such views can be grouped under "monistic idealism." Monistic idealism maintains that consciousness, not matter, is the ground of all being. It is monist because there is only one substance in the Universe: consciousness. In a sense, the journey has come full circle, back to the original question, "What is consciousness?"

To a monistic idealist, consciousness is both material and non-material - mind and matter. There is no mind without matter and no matter without mind. Monistic idealism conceptualizes the meshing of opposites. There is no male without female, no up without down, no existence without non-existence. Consciousness is the unification of the internal with the external. The relationship between monistic idealism and quantum mechanics is connected to the idea of "wave-particle duality" revealed through the "consciousness causes the collapse of the wave function" interpretation of quantum mechanics. According to monistic idealism, everything in the Universe is connected by a vast oneness (A. Goswami (1995)) - a consciousness of everything until that consciousness becomes aware of something. At that point, a particular outcome or event occurs. I become aware of that pencil in front of me. The Eastern mystic aspires to become unaware of particular things and become aware of only consciousness itself, which is the supreme spiritual epiphany.

As it pertains to the two philosophical questions, according to monistic idealism, an external world depends on the existence of an internal world. There can be no separation of the two. The mind and the body are inseparable. Awareness of a body without a mind is impossible. Hence, our consciousness is aware of having a body - a specific event or outcome in this vast landscape of interconnected consciousness.

In general, physicists have looked at monistic idealism skeptically, primarily because it claims to have solved the measurement problem in quantum mechanics. And while there are many proposed solutions to the measurement problem, no solution has gained general acceptance. Monistic idealism is not very scientific in this sense. It claims that what exists is nothing but consciousness and yet does not explain the mechanism of how that consciousness arises. If the mechanism cannot be explained, on what grounds can it be said that consciousness is understood? To put it another way, how did conscious beings arise out of the big bang? Moreover, most quantum physicists would deny that pure consciousness could be known beyond assigning probabilities to outcomes. If it is true that consciousness entails both the physical and the mental, all in one package, that package cannot be further subdivided into distinct parts.

# Epilogue

This concludes a lengthy discussion of modern physics, its history, and the logical and philosophical implications of attempting to give a unified view of our world. What did we learn? Two grand theories describe the nature of our Universe. General relativity describes the physics of the macro-Universe, and quantum mechanics the physics of the micro-Universe. The two theories explain the origins of almost everything, but not quite.

The standard model of particle physics successfully explains three of the four fundamental interactions of Nature. But it also suffers several shortcomings. First, it requires nineteen or so adjustable constants whose values lack a theoretical explanation. Second, it does not include gravity or explain the origins of dark energy/matter. Third, it does not accommodate the observed predominance of matter over antimatter. Fourth, the current theory requires that neutrinos have zero mass which does not agree with recent experiments.

So that its predictions agree with observations, the model must be supplemented with unnatural mechanisms, renormalization and the Higgs mechanism, for example, which are left theoretically unexplained. In addition, the input parameters to the model must be unnaturally fine-tuned so that the model predictions agree with observations. Finally, the self-consistency of the theoretical part of the model has yet to be mathematically demonstrated (the mass gap).

On the other hand, while the theory of relativity successfully describes the macro part of our Universe, and while it does not suffer serious logical or mathematical foundational issues, it cannot explain micro-level physics. To apply general relativity to small scales requires that its nonlinear equations be quantized in some manner. But, all attempts to do so have resulted in non-renormalizable theories - theories whose infinities cannot be discarded or subsumed. This fact has driven most physicists to the safer harbingers of string theory.

In most problems with which physicists deal, the dichotomy between the theory of relativity and the quantum theory is irrelevant. For example, relativists can study the cosmos without worrying much about quantum effects. And quantum physicists can study the microworld without worrying about the effects of gravity. But there are phenomena, black holes, for example, that cannot be explained by employing one theory or the other. In these situations, the incompatibilities between the two theories become critical.

So far, all attempts at unifying the two theories into a quantum theory of gravity have failed. In Chapter 4, we uncovered what appears to be the root of the problem. The logical foundations of the two theories are vastly different.

Some would argue that the fundamental differences between the two theories will never be overcome - that the correct approach is to relinquish the belief that a unified theory will materialize. Accordingly, physicists must concede that physics will consist of a plurality of theories that depend on the particular phenomenon under consideration.

Still, others hope that string theory will remove the inconsistencies between relativity and quantum mechanics without deviating too much from already established physics. But which string theory? Will M-theory provide a *"theory of everything,"* or will there be a landscape of string theories, where only one of many theories in a vast array of universes will describe the nature of our Universe or our part of the Universe?

Finally, is there a different approach to unification - a road yet traveled? Finding a unified physics theory will require the logical inconsistencies in current theories to be understood and overcome. In addition, the philosophical issues raised in Chapter 5 must be addressed. On the face of it, it appears that the road to everywhere leads to nowhere. So, the question, "Will there be a unified theory of the world?" remains open.

# List of References

B. De Spinoza (1677) [1949], *Ethics preceded by on the improvement of the understanding,* J. Gutmann, Ed., Hafner Publishing Co., NY.

E. Noether (1918), Nachrichten der Königlichen Gesellschaft der Wissenschaften zu Göttingen, *Math.-phys.* Klasse, pp. 235-257.

T. Kaluza (1921), Zum unitätsproblem in der physic, *Sitzungsber, Preuss. Akad. Wiss,* Berlin. (Math. Phys.): pp. 966-972.

W. Gerlach, O. Stern (1922), "Der experimentelle nachweis der richtungsquantelung im Magnetfeld. *Zeitschrift für Physik. 9 (1)*: pp, 349-352. Bibcode:1922ZPhy....9..349G. doi:10.1007/BF01326983. S2CID 186228677.

O. Klein (1926), Quantentheorie und fünfdimensionale relativitätstheorie, *Zeitschrift für Physik A 37 (12):* pp. 895-906. Bibcode 1926ZPhy...37..895K. doi:10.1007/BF01397481.

E. Schrödinger (1926), An undulatory theory of the mechanics of atoms and molecules, *Physical Review. 28 (6):* pp. 1049-1070. Bibcode:1926PhRv...28.1049S.

G. Uhlenbeck, S. Goudsmit (1926), Spinning Electrons and the Structure of Spectra, *Nature 117,* pp. 264-265, https://doi.org/10.1038/117264a0.

W. Heisenberg (1927), Über den anschaulichen inhalt der quantentheoretischen kinematik und mechanic, *Z. Phys. 43 (3-4)*: pp. 172-198. Bibcode:1927ZPhy...43..172H. doi:10.1007/BF01397280. S2CID 122763326.

G. Gamow (1928), Zur quantentheorie des atomkernes, *Z. Physik 51,* 204. 65.

A. Einstein, B. Podolsky, N. Rosen (1935). Can quantum-mechanical description of physical reality be considered complete?, *Physical Review 47 (10)*: pp. 777–780. Bibcode 1935PhRv...47..777E. doi:10.1103/PhysRev.47.777.

H. Weyl (1950) [1931], *The theory of groups and quantum mechanics,* Translated by H. P. Robertson, Dover Publications, NY.

D. Bohm (1951), *The quantum theory,* Dover Publications, Inc., NY.

C. N. Yang and R. L. Mills (1954), Conservation of isotopic spin and isotopic gauge invariance, *Phys. Rev. 96,* 191.

A. Einstein (1956), *Investigations on the theory of the Brownian movement,* edited by R. Fürth, Translated by A.D. Cowper, Dover Publications Inc, NY.

H. Everett (1957), Relative state formulation of quantum mechanics, *Reviews of Modern Physics 29,* 454.

M. Gell-Mann (1961), The eightfold way: a theory of strong interaction symmetry, *Synchroton Laboratory Report CTSL-20* (California Institute of Technology).

L. Wittgenstein (1961) [1921], *Tractatus Logico-Philosophicus,* Translation by D. F. Pears and B. F. McGuinness, Routledge & Kegan Paul, UK.

P. W. Anderson (1962), Plasmons, gauge invariance, and mass, *Physical Review 130 (1)*: pp. 439–442. Bibcode 1963PhRv..130..439A. doi:10.1103/PhysRev.130.439.

G. J. Warnock (1963), *Introduction to: the principles of human knowledge and three dialogues between Hylas and philonous by G. Berkeley,* Meridian Books, OH.

B. Blanshard (1964), *Reason and analysis,* Open Court Publishing Co, IL.

B. J. Bjorken, S. L. Glashow (1964), Elementary particles and SU(4), *Physics Letters 11 (3)*: pp. 255–257.

F. Englert and R. Brout (1964), Broken symmetry and the mass of gauge vector mesons, *Physical Review Letters 13 (9)*: 321–323. Bibcode 1964PhRvL..16..321E.

M. Gell-Mann (1964), A schematic model of baryons and mesons, *Physics Letters 8 (3)*: 214–215. Bibcode 1964 PhL.....8..214G. doi:10.1016/S0031-9163(64)92001-3.

R. P. Feynman, R. B. Leighton and M. Sands (1964), *The Feynman lectures on physics,* California Institute of Technology.

P. W. Higgs (1964), Broken symmetries and the masses of gauge bosons, *Physical Review Letters 13 (16)*: 508–509. Bibcode 1964PhRvL..16..508H. doi:10.1103/PhysRevLett.16.508.

R. Penrose (1965), Gravitational collapse and space-time singularities, *Phys. Rev. Lett. 14,* 57.

A. Beiser (1967), *Concepts of modern physics,* McGraw-Hill, NY.

R. Descartes (1968), *The meditations and selections from the principles of René Descartes,* Translated by J. Veitch, The Open Court Library of Philosophy, Chicago.

D. Hume (1968), *A treatise of human understanding,* Reprinted from the Original Edition of Three Volumes, L. A. Selby-Bigge, Ed., Oxford at the Clarendon Press.

M. Breidenbach (1969), Observed behavior of highly inelastic electron-proton scattering, *Physical Review Letters 23 (16)*: 935–939. Bibcode 1969PhRvL..23..935B.

R. Haag, J. T. Lopuszanski and M. Sohnius (1975), All possible generators of supersymmetries of the s matrix, *Nucl. Phys. B 88* 257.

J. S. Bell (1981), *Quantum mechanics for cosmologists,* In C.J. Isham, R. Penrose and D.W. Sciama (eds.), Quantum Gravity 2: A second Oxford Symposium. Oxford: Clarendon Press, p. 611.

D. Griffiths (1987), *Introduction to elementary particles,* John Wiley & Sons., pp. 280–281. ISBN 0-471-60386-4.

A. Pickering (1984), *Constructing quarks,* University of Chicago Press. pp. 114–125. ISBN 0-226-66799-5.

G. Ross (1984), *Grand unified theories,* Westview Press, CO. ISBN 978-0-8053-6968-7.

H. C. Ohanian (1986), What is spin?, *American Journal of Physics 54,* 500; https://doi.org/10.1119/1.14580.

J. S. Bell (1987), On the Einstein Podolsky Rosen paradox, *Physics Physique Физика. 1 (3):* 195–200, doi:10.1103/PhysicsPhysiqueFizika.1.195.

B. Greene, J. Schwarz and E. Witten (1987), *Superstring theory, volumes 1 and 2,* Cambridge UP, Cambridge MA.

S. Weinberg (1987), Anthropic bound on the cosmological constant, *Phys. Rev. Lett. 59 (22)*: 2607–2610. Bibcode 1987PhRvL..59.2607W. doi:10.1103/PhysRevLett.59.2607. PMID 10035596.

D. Halliday and R. Resnick (1988), *Fundamentals of physics,* John Wiley & Sons, NY.

B. Hatfield (1989), *Quantum field theory of point particles and strings,* Addison-Wesley, Reading, MA.

P. J. Nahin (1992), Maxwell's grand unification, *Spectrum, IEEE 29 (3):* 45. doi:10.1109/6.123329.

D. Bohm & B. J. Hiley (1993), *The undivided universe: an ontological interpretation of quantum theory,* Routledge, NY.

D. Lindley (1993), *The end of physics: the Myth of a unified field theory,* Basic Books, NY.

S. Weinberg (1993), *Dreams of a final theory: the scientist's search for the ultimate laws of nature,* Vintage Books, NY, pp. 153-160.

C. M. Hull and P. K. Townsend (1994), Unity of superstring dualities, *QMW-94-30 R/94/33 hep-th/9410167.*

C. B. Parker (1994), *McGraw Hill Encyclopedia of Physics (2nd ed.).* McGraw Hill. p. 541. ISBN 0-07-051400-3.

B. Carithers, P. Grannis (1995), Discovery of the top quark, *Beam Line (SLAC) 25 (3)*: pp. 4–16. Retrieved 2008-09-23.

A. Goswami (1995), *The self-aware universe: how consciousness creates the material world,* Jeremy P. Tarcher/Putnam, NY.

J. Polchinski (1995), An article which established D-branes' significance in string theory, *Phys. Rev. Lett. 75,* 4724.

Z. Schreiber (1995), *The nine lives of Schrödinger's cat: on the interpretation of non-relativistic quantum mechanics,* Rakach Institute of Physics, The Herbrew University Givaat Ram Jerusalem.

L. Susskind (1995), The world as a hologram, *Journal of Mathematical Physics 36 (11)*: 6377–6396. arXiv:hep-th/9409089. Bibcode 1995JMP....36.6377S. doi:10.1063/1.531249.

S. I. Tomonaga (1997), *The story of spin,* University of Chicago Press. p. 35. ISBN 0-226-80794-0.

Y. Fukuda (1998), Measurements of the solar neutrino flux from super-Kamiokande's first 300 Days, *Physical Review Letters 81 (6)*: 1158–1162. arXiv:hep-ex/9805021. Bibcode 1998PhRvL..81.1158F. doi:10.1103/PhysRevLett.81.1158.

D. Gross, V. Periwal (1998), Gross and Periwal reply, *Phys. Rev. Lett. 61,* 1517.

I. Newton (1998), *Opticks,* Commentary by Nicholas Humez (Octavo ed.). Palo Alto, Calif.: Octavo. ISBN 1-891788-04-3.

J. Polchinski (1998), *String theory, volumes 1 and 2,* Cambridge UP, Cambridge MA.

A. Aspec (1999), Bell's inequality test: more ideal than ever, *Nature 398 189.*

B. Greene (1999), *The elegant universe: superstrings, hidden dimensions, and the quest for the ultimate theory,* Vintage Books, NY.

D. M. Harrison (1999), *Bell's theorem,* Department of Physics, University of Toronto, harrison@physics.utoronto.ca.

E. Joos (2000), *Decoherence: theoretical, experimental, and conceptual problems*", edited by P. Blanchard, D. Giulini, E. Joos, C. Kiefer, and I.O. Stamatescu (Springer, Berlin), pp. 1–17.

H. Kragh (2000), *Max Planck: the reluctant revolutionary,* Physics World, https://physicsworld.com/a/max-planck-the-reluctant-revolutionary/.

R. B. Laughlin, D. Pines (2000), The theory of everything, *The National Academy of Sciences.*

S. E. Rugh, H. Zinkernagel (2000), The quantum vacuum and the cosmological constant problem, *Studies in History and Philosophy of Modern Physics, vol. 33,* 663-705: arXiv:hep-th/0012253v1.

C. Callender, N. Huggett (2001), *Physics meets philosophy at the Planck scale,* Cambridge University Press, NY.

D. M. Harrison (2001), *The Stern-Gerlach experiment, electron spin, and correlation Experiments,* Dept. of Physics, University of Toronto, Canada.

S. Rugh (2001), The quantum vacuum and the cosmological constant problem, *Studies in History and Philosophy of Modern Physics 33 (4)*: pp. 663–705. doi:10.1016/S1355-2198(02)00033-3.

D. D. Vvedensky (2001), *Group theory,* Lecture notes.

E. Witten (2002), *Can scientists' "theory of everything" really explain all the weirdness the universe displays?,* www.astronomy.com.

J. Greensite (2003), *Lecture notes on quantum mechanics,* Physics and Astronomy Department, San Francisco State University, CA.

C. T. Hill, E. H. Simmons (2003), Strong dynamics and electroweak symmetry breaking, *Physics Reports 381 (4-6)*: pp. 235–402, arXiv:hep-ph/0203079. Bibcode 2003PhR...381..235H. doi:10.1016/S0370-1573(03)00140-6.

L. Susskind (2003), *The anthropic landscape of string theory,* Department of Physics Stanford University, Stanford, CA 94305-4060.

H. F. M. Goenner (2004), On the History of Unified Field Theories, *Living Rev. Relativ. 7, 2.* https://doi.org/10.12942/lrr-2004-2S.

S. Singh (2004), *Big bang: the origin of the universe,* Harper-Perennial, NY.

A. Ashtekar (2005), Gravity and the quantum, *New Journal of Physics 7* 198.

D. Gross (2005), Einstein and the search for unification, *CURRENT SCIENCE, VOL. 89, NO. 12, 25 DECEMBER 2005.*

M. Isaak (2005), *CI301: The anthropic principle,* Index to Creationist Claims, TalkOrigins Archive. Retrieved 2007-10-31.

G. P. Lepage (2005), *What is renormalization?,* Newman Laboratory of Nuclear Studies, Cornell University, Ithaca, NY 14853, Talk presented at TASI'89, June 1989.

R. Penrose (2005), *The road to reality: a complete guide to the laws of the universe,* Alfred A. Knopf, NY.

M. Schlosshauer (2005), *Decoherence, the measurement problem, and interpretations of quantum mechanics,* Department of Physics, University of Washington, Seattle, Washington 98195, USA.

S. Weinberg (2005), *Einstein's mistakes,* Physics Today, November 2005, page 31.

S. Battersby (2006), *Introduction: cosmology,* NewScientist.

M. P. Hobson, G. P. Efstathiou, A. N. Lasenby (2006), *General relativity: an introduction for physicists* (Reprinted with corrections 2007 ed.), Cambridge University Press. p. 187. ISBN 9780521829519.

R. Oerter (2006), *The theory of almost everything: the standard model, the unsung triumph of modern physics* (p. 2), Penguin Group. Kindle Edition.

P. Woit (2006), *Not even wrong: the failure of string theory and the search for unity in physical law,* Perseus Books, NY.

P. Byrne (2007), *The many worlds of Hugh,* Scientific American, Dec 2007.

L. Smolin (2007), *The trouble with physics: the rise of string theory, the fall of science and what comes Next,* Houghton Mifflin Co., Boston, MA.

T. Sauer (2007), *Einstein's unified field theory program,* Einstein Papers Project, California Institute of Technology 20-7, Pasadena, CA 91125, USA.

G. Giudice (2008), Naturally speaking: the naturalness criterion and physics at the LHC. *Perspectives on LHC Physics.* pp. 155–178.

P. Rincon (2008), *Fermilab "God Particle" may have been seen,* Retrieved on 2008-05-27".

G. Rodreguez-Peretra (2008), Descartes's substance dualism and his independence conception of substance, *Journal of the History of Philosophy 46:1.*

M. Kumar (2009), *Quantum,* Icon Books, London, p. 305.

J. M. Marin (2009), Mysticism in quantum mechanics: the forgotten controversy, *European Journal of Physics 30*: 807–822.

B. R. Martin, G. Shaw (2009), *Particle physics (3rd Edition),* Manchester Physics Series, John Wiley & Sons, NY, ISBN 978-0-470-03294-7.

D. McMahon (2009), *String theory demystified: a self-teaching guide,* McGraw-Hill, NY.

D. Tong (2009), *String Theory,* University of Cambridge Part III Mathematical Tripos, Department of Applied Mathematics and Theoretical Physics, Centre for Mathematical Sciences, Wilberforce Road, Cambridge, CB3 OWA, UK.

L. ´Alvarez-Gaum´e, M. A. V´azquez-Mozob (2010), *Introductory lectures on quantum field theory,* Physics Department, Theory Division, CERN, CH-1211 Geneva 23, Switzerland.

A. Bokulich (2010), *Bohr's correspondence principle,* The Stanford Encyclopedia of Philosophy (Winter 2010 Edition), Edward N. Zalta (ed.), URL = <http://plato.stanford.edu/archives/win2010/entries/bohr-correspondence/>.

P. Pearle, B. Collett, K. Bart, D. Bilderback, D. Newman and S. Samuel (2010), What Brown saw and you can too, *Am. J. Phys. 78*: pp. 1278-1289.

A. Z. Jones, D. Robbins (2010), *String theory for dummies,* Wiley Publishing, Inc., NJ.

M. Rohlf (2010), *Immanuel Kant,* The Stanford Encyclopedia of Philosophy (Fall 2010 Edition), Edward N. Zalta (ed.), URL = <http://plato.stanford.edu/archives/fall2010/entries/kant/>.

A. Zee (2010), *Quantum field theory in a nutshell,* Princeton University Press, NJ.

J. S. Bell (2011), *La nouvelle cuisine,* last article of Speakable and Unspeakable in Quantum Mechanics, second edition. Cambridge University Press, UK.

D. Metzler (2011), *Million dollar math and physics: the Yang-Mills problem or: the Jekyll and Hyde story about the standard model,* YouTube.

S. Weinstein, D. Rickles (2021), *Quantum gravity,* The Stanford Encyclopedia of Philosophy (Fall 2021 Edition), Edward N. Zalta (ed.), URL = <https://plato.stanford.edu/archives/fall2021/entries/quantum-gravity/>.

D. Castelvecchi (2012), *Is supersymmetry dead?,* Scientific American.

A. Goshaw (2012), *Physics 346 – a course in the standard model,* Duke University, NC.

R. D. Klauber (2012), *Student friendly quantum field theory,* Sandtrove Press, IA.

J. Schombert (2012), *HC 209: 21st century science,* Prof. Office: 461 Willamette, 6-5214, University of Oregon.

J. R. Taylor, C. D. Zafiratos, M. A. Dubson (2015), *Modern physics for scientists and engineers (2nd ed.),* University Science Books, pp. 136–9.

C. Sutton (2020), *Unified field theory*, Encyclopedia Britannica, 28 Feb. 2020, https://www.britannica.com/science/unified-field-theory. Accessed 6 May 2023.

A. Jaffe, E. Witten (N/A), *Quantum Yang-Mills theory,* Official problem description.

Made in the USA
Monee, IL
04 November 2026

67387543R00115